Number 290

Baruch Solel

Irreducible triangular algebras

Memoirs

of the American Mathematical Society

Providence · Rhode Island · USA

January 1984 · Volume 47 · Number 290 (second of six numbers) · ISSN 0065-9266

Memoirs of the American Mathematical Society

Number 290

Baruch Solel

Irreducible triangular algebras

Published by the
AMERICAN MATHEMATICAL SOCIETY
Providence, Rhode Island, USA

January 1984 · Volume 47 · Number 290 (seco six numbers)

MEMOIRS of the American Mathematical Society

This journal is designed particularly for long research papers (and groups of cognate papers) in pure and applied mathematics. It includes, in general, longer papers than those in the TRANSACTIONS.

Mathematical papers intended for publication in the Memoirs should be addressed to one of the editors. Subjects, and the editors associated with them, follow:

MEMOIRS are printed by photo-offset from camera-ready copy fully prepared by the authors. Prospective authors are encouraged to request booklet giving detailed instructions regarding reproduction copy. Write to Editorial Office, American Mathematical Society, P. O. Box 6248, Providence, Rhode Island 02940. For general instructions, see last page of Memoir.

SUBSCRIPTION INFORMATION. The 1984 subscription begins with Number 289 and consists of six mailings, each containing one or more numbers. Subscription prices for 1984 are $148 list; $74 member. A late charge of 10% of the subscription price will be imposed upon orders received from nonmembers after January 1 of the subscription year. Subscribers outside the United States and India must pay a postage surcharge of $10; subscribers in India must pay a postage surcharge of $15. Each number may be ordered separately; *please specify number* when ordering an individual number. For prices and titles of recently released numbers, refer to the New Publications sections of the NOTICES of the American Mathematical Society.

BACK NUMBER INFORMATION. For back issues see the AMS Catalogue of Publications.

TRANSACTIONS of the American Mathematical Society

This journal consists of shorter tracts which are of the same general character as the papers published in the MEMOIRS. The editorial committee is identical with that for the MEMOIRS so that papers intended for publication in this series should be addressed to one of the editors listed above.

Subscriptions and orders for publications of the American Mathematical Society should be addressed to American Mathematical Society, P. O. Box 1571, Annex Station, Providence, R. I. 02901. *All orders must be accompanied by payment.* Other correspondence should be addressed to P. O. Box 6248, Providence, R. I. 02940.

MEMOIRS of the American Mathematical Society (ISSN 0065-9266) is published bimonthly (each volume consisting usually of more than one number) by the American Mathematical Society at 201 Charles Street, Providence, Rhode Island 02904. Second Class postage paid at Providence, Rhode Island 02940. Postmaster: Send address changes to Memoirs of the American Mathematical Society, American Mathematical Society, P. O. Box 6248, Providence, RI 02940.

TABLE OF CONTENTS

ABSTRACT

This memoir is devoted to the study of the structure of irreducible triangular algebras generated by a maximal abelian algebra and an ordered semigroup of unitary operators acting on the maximal abelian algebra.

1980 Mathematics Subject Classification. Primary 47D25, Secondary 46L99, 47A35.

Key words and pharases : Triangular algebra, irreducible algebra, action of a unitary group, Arveson's expectation, automorphism, derivation.

Library of Congress Cataloging in Publication Data

Solel, Baruch, 1952-
Irreducible triangular algebras.

(Memoirs of the American Mathematical Society, ISSN 0065-9266 ; no. 290)
"Volume 47, number 290 (second of six numbers)"
Bibliography: p.
1. Triangular operator algebras. 2. Operator algebras. 3. Ergodic theory. I. Title. II. Series.
QA3.A57 no. 290 [QA326] 510s [512'.55] 83-22524
ISBN 0-8218-2290-X

INTRODUCTION

This memoir is devoted to the study of the structure of irreducible triangular algebras generated by a maximal abelian algebra $\mathcal{a}$ and an ordered semigroup G of unitary operators acting on $\mathcal{a}$. In its basic mathematical aspects, this investigation follows two paths. Along the first, and most primitive, it is a further development of the structure theory of a subclass of non-self-adjoint operator algebras - the irreducible triangular algebras. Along the second, it is an exploration of some parts of non-commutative ergodic theory - with emphasis on non-self adjoint features of this theory.

We elaborate somewhat on both these aspects of our investigation. The theory of triangular operator algebras was initiated by Kadison and Singer in a paper [33] which appeared in 1960. With H a complex Hilbert space and B(H) the algebra of all bounded operators on it, a subalgebra $\mathcal{J}$ of B(H) such that $\mathcal{J} \cap \mathcal{J}^*$ is maximal abelian in B(H) is said to be <u>triangular</u> and $\mathcal{J}\cap\mathcal{J}^*$ is said to be its <u>diagonal</u>. An application of Zorn's lemma permits us to conclude that $\mathcal{J}$ is contained in a maximal triangular algebra $\mathcal{J}_0$ (and, necessarily, $\mathcal{J} \cap \mathcal{J}^* = \mathcal{J}_0 \cap \mathcal{J}_0^*$). The projection E is said to be a hull (in $\mathcal{J}$) when E is <u>invariant</u> under each T in $\mathcal{J}$ - that is ETE=TE for each T in $\mathcal{J}$ (the range of E is <u>stable</u> under each T in $\mathcal{J}$). Such a projection, E, is invariant under the self-adjoint algebra $\mathcal{J}\cap\mathcal{J}^*$, hence commutes with $\mathcal{J} \cap \mathcal{J}^*$ - and, hence, lies in (the maximal abelian) $\mathcal{J} \cap \mathcal{J}^*$.

If H is finite dimensional, an (ordered) orthonormal basis $e_1,\dots,e_n$ for H can be chosen so that $\mathcal{J} \cap \mathcal{J}^*$ appears as the algebra of all diagonal matrices relative to this basis and $\mathcal{J}$ appears as the algebra of (upper) triangular matrices relative to this (ordered) basis. Again, if H is infinite dimensional and $\mathcal{J} \cap \mathcal{J}^*$ is generated by its minimal (that is, one-dimensional) projections (we say that $\mathcal{J} \cap \mathcal{J}^*$ is <u>totally-atomic</u>, in this case),

Received by the editors June 8, 1982.

we can choose an orthonormal basis and an ordering on it such that $\mathcal{g}$ appears as the algebra of all bounded operators whose matrices are upper triangular relative to the basis in that ordering. (see [33, Theorem 3.2.1].) For a finite basis the (total) ordering is not essential - there is only one total ordering - up to order isomorphism. In the infinite-dimensional case the (total) ordering is essential (the basis might be ordered as the rationals, or as the integers, with a first or last element or without, among other possibilities); and it is shown in [33, Theorem 3.2.1] that two maximal triangular algebras with totally-atomic diagonals are unitarily equivalent if and only if their sets of minimal projections are order isomorphic.

For the more general triangular algebra $\mathcal{g}$, it is still the case that the set of hulls is totally ordered (by the usual projection ordering). (See [33, Lemma 2.3.3].) But it need not be true that there are very many hulls. The (abelian) von Neumann subalgebra C of $\mathcal{g} \cap \mathcal{g}^*$, generated by the hulls, is called the core of $\mathcal{g}$. In the case of a totally-atomic diagonal, $C = \mathcal{g} \cap \mathcal{g}^*$. The maximal triangular algebras for which the core coincides with the diagonal are called hyperreducible by Kadison and Singer; and this class is the one analyzed in detail in [33]. The diagonal need not be totally atomic if the algebra is hyperreducilbe. In fact there is one whose diagonal contains no minimal projections (and only one, up to unitary equivalence, in the case of separable Hilbert space -see [33, Theorem 3.3.1].) The opposite extreme is the case where C consists of scalar multiples of the identity operator I. Such algebras are called irreducible in [33], and examples are given (based on [33, Theorem 2.2.1]) to show that the class of such algebras is not empty.

An orthonormal basis in a Hilbert space gives rise to a totally-atomic maximal abelian algebra and such an albegra corresponds to an (essentially unique) orthornormal basis. It is natural to regard the general maximal abelian algebra as a generalized orthonormal basis. In the same way, a maximal triangular algebra with totally-atomic diagonal corresponds to an ordered orthonormal basis; and a general hyperreducible maximal triangular algebra with totally-atomic diagonal corresponds to an ordered orthonormal basis; and a general hyperreducible maximal triangular algebra should be viewed as a generalized, oredered orthonormal basis.

Part of the motivation for the study of the maximal triangular algebras lies in their role in formalizing the concept of a generalized ordered orthonormal basis. Another part of the motivation lies in the general attempt to put arbitrary bounded operators on a Hilbert space in "triangular form" and, more broadly, to reduce certain algebras of operators to triangular form. For this purpose, the hyperreducible algebras provide the most satisfactory "triangularization".

The irreducible triangular algebras appear as a byproduct of the studies related to the first two motivating aspects of the total investigation. In some sense, they provide the key to the third motivating aspect. As noted in the introduction to [33], Kadison and Singer expect the (maximal) triangular algebras to serve in the theory of non-self-adjoint operator algebras in something resembling the position that the von-Neumann algebras fill in the self-adjoint theory. Carrying this correspondence further, we would associate the hyeperreducible algebras with the abelian von Neumann algebras. (The core of a triangular algebra would correspond to the center of the von Neumann algebra.) The irreducible maximal triangular algebras would be associated with the factors (in the theory of von Neumann algebras). As a consequence, we might expect the problems associated with the irreducible triangular algebras to be the most intractable. This turned out to be the case. The questions concerning this class of triangular algebras posed in [33] remain open. (See [33, Qestions 2.4.4, 2.4.5, 2.4.6].)

We extend the construction of Kadison and Singer [33, Theorem 2.2.1, Example 2.2.2] to include (free, ergodic) actions of ordered semigroups G of unitary operators on a maximal abelian algebra $\mathcal{a}$. We prove the extended version of [33, Theorem 2.2.1], the key point of which establishes the "$\mathcal{a}$-independence of G" and of $\bar{G}$, the group generated by G. (See Proposition

1.2 and Corollary 1.3.) That construction provides us with the class of irreducible triangular algebras whose detailed analysis is the subject of this treatise. The measure-theoretic constructs and results needed for this construction and the specific examples are contained in Chapter I. Along the lines of the structure theory of the irreducible triangular algebras our examples provide us with instances of such algebras that are not isomorphic. (See Corollary 3.24 and Example 3.25.) This is another step in the direction of answering Question 2.4.6 of [33]: Are there two (or more) irreducible maximal triangular algebras on separable Hilbert space which are not algebraically isomorphic. The first such step was taken by Arveson and Josephson in [7, Theorem 4.1] where they distinguish such (norm-closed) algebras in terms of conjugacy classes of ergodic measure preserving transformations. Our techniques allow us to decide when the acting semigroup is singly generated. Although the fundamental difficulty remains of describing the operators in some irreducible maximal triangular algebra (in manner susceptible to computational application), these maximal algebras are reached, in the final stage, by an application of Zorn's Lemma; it seems well worthwhile to bypass this difficulty and study explicitly constructed classes of irreducible triangular algebras.

The second path we follow, that of non-commutative (non-self-adjoint) ergodic theory, relates to the fact that the class of examples we study arises from constructions in ergodic theory.
The origin of operator algebra considerations in ergodic theory is, of course, the "group-measure space" construction of Murray and von Neumann [41, pp. 192-209] with which they produced their first examples of type II factors. In two papers, basic to this development [12,13], Dye carries the study of groups of measure-preserving transformations further, introducing several constructs and techniques that are key aspects of the theory. In particular, a variant and elaboration of his "F-analysis" is developed in the second section of our Chapter II. Combining it with estimates (Theorem 2.14 and Lemma 2.15), this

analysis allows us to detect (normalizing) unitary operators in the algebra generated by the (ordered) semigroup and the maximal abelian algebra on which it acts (as distinguished from its various closures - cf. Theorem 2.16 and Corollary 2.17) by their action on the maximal abelian algebra.

In an important sense, from the ergodic-theoretic viewpoint, our study is a direct descendant of the work of Arveson in this area [3,7]. One of the vital tools in our investigation is the "Arveson expectation". (Lemma 2.1 through 2.9 are a recapitulation of Arveson's results related to this expectation.) With regard to the relation between [3,7] and the results of this study, we note that Arveson's emphasis is on relating the algebraic structure of the (norm-closed) algebra (and C*-algebra) generated by the action of a single transformation (a discussion of more general groups appears in [3, Section 2]) to the (measure) conjugacy class of the transformation; while our emphasis is on passing from these algebras, generated by ordered semigroups, to the structure of the semigroups (and associated groups). Samples of results of this nature are to be found in Proposition 2.27, Corollary 2.28, Theorem 2.29 and Proposition 2.30.

In connection with the general subject of triangular operator algebras, important work following [33] is to be found in Arveson [1], Erdos [14], Hopenwasser [31], Rosental [46] and Schue [48].
Ringrose extended the study of the hyperreducible algebras to "nest algebras" [44] and these were further developed by Christensen [9], Deddens [10], Erdos [15,16,17,18,19], Fall, Arveson and Muhly [20],Gilfeather and Larson [23,24, 25], Larson [36], Longstaff [37], Lance [34,35], Ringrose [43,45] and Arveson [6].
The reflexive algebras were introduced and studied by Halmos [26] and further studied by Arveson [4], Hopenwasser [30,32], and Longstaff [38,39,40].

We acknowledge with gratitude several conversations with Professor Effros at the earliest stage of our interest in this erea of operator-ergodic theory.

Our thanks are due to Professor Kadison for crystallizing the specific topic of this thesis, for his patient guidance of our research and for his careful reading of the thesis and many helpful suggestions.
Professor Arveson kindly filled in for us the details of an arguement alluded to in [2] which enabled us to include Example 4.25. We are grateful for this help. Our scientific debts to the work of Arveson [3.7] , Dye [12,13] and Kadison-Singer [33] will be apparent to any reader familiar with those papers.

We described, briefly, some of the major points in the four chapters that follow.

In Chapter I we prove that if G is an ordered semigroup of unitary operators acting freely and ergodically on a maximal abelian algebra $\mathcal{A}$ then the algebra $\mathcal{G}$, generated by $\mathcal{A}$ and G, is an irreducible triangular algebra. The study of this algebra will be taken up in the following chapters.
Several examples are presented, among them is the algebra generated by the multiplication algebra on $L^2(\mathbb{T})$ ($\mathbb{T}$ is the unit circle) and the unitary operator associated with an irrational rotation (this example appeared in [33]).

For another example we construct an ordered semigroup T of real numbers, dense in $\mathbb{R}$, and show that the semigroup of the unitary operators on $L^2(\mathbb{R})$ associated with translations of $\mathbb{R}$ by members of T generates, with the multiplication algebra on $L^2(\mathbb{R})$, an irreducible triangular algebra.

Chapter II studies the structure of an irreducible triangular algebra generated by a maximal abelian algebra $\mathcal{G}$ and a semigroup G of unitary operators acting freely and ergodically on $\mathcal{A}$.

We present the Arveson's expectation from the C*-algebra $\mathcal{B}(\mathcal{A},G)$ generated by $\mathcal{G}$ onto $\mathcal{A}$ (see [3]) and some of its properties.
We note that if the group $\bar{G}$, generated by G, is amenable the expectation is faithful (Proposition 2.7). Under a condition (that holds for a semigroup

of unitary operators arising from measure preserving transformations on a finite measure space) the converse is also true.

We discuss unitary operators acting on $\mathcal{a}$ and define, for such an operator W, a projection F(W) in $\mathcal{a}$ indicating, roughly, the part of the Hilbert space H on which W acts "locally" as an operator in $\bar{G}$. This definition is inspired by Dye's analysis of groups of measure preserving automorphisms (see [12]) and so is the following result (see Theorem 2.16): If W is a unitary operator acting on $\mathcal{a}$ then F(W)=I if and only if $W = \Sigma_{U \in \bar{G}} VP_U U$ where the sum is understood in the sense of strong-operator convergence, $\{P_U\}$, $\{U^*P_U U\}$ are orthogonal families of projections in each with sum I, and V is a unitary operator in $\mathcal{a}$.
We use this analysis and some estimates (Theorem 2.14 and Lemma 2.15) to prove (Corollary 2.17) that each unitary operator in $\mathcal{B}(\mathcal{a},G)$, acting on $\mathcal{a}$, is actually in $g+g^*$.

Next, we investigate the relationship between the order properties of G and the structure of g. For example, we find (Theorem 2.29) that if G_1 and G_2 are two ordered semigroups acting freely and ergodically on the abelian algebra $\mathcal{a}$ and generating the same irreducible triangular algebra g then G_1 is generated by a single operator if and only if G_2 is generated by a single operator. Moreover, if G_1 is generated by U and G_2 by V then $UV^* \in \mathcal{a}$

In Chapter III we define a homomorphism from the irreducible triangular algebra g into B(K), for some Hilbert space K, to be a linear multiplicative map whose restriction to $\mathcal{a}$ is self adjoint.

For a homomorphism that maps $\mathcal{a}$ onto a maximal abelian algebra we find conditions that are equivalent to the homomorphism being an isomorphism (Proposition 3.50). For example, we show that if the homomorphism maps each unitary operator in G into an invertible operator and if its restriction to $\mathcal{a}$ ia an isomorphism, then the homomorphism is an isomorphism.

We discuss the possibility of extending an automorphism on g, that maps $\mathcal{a}$ onto itself, to a *-automorphism on $g+g^*$ and, furthermore, to a

-automorphism on $\mathcal{B}(\mathcal{a},G)$ (the C-algebra generated by $\mathcal{g}$).
We see (Corollary 3.14) that if Ψ is an automorphism of $\mathcal{g}$ that maps $\mathcal{a}$ onto itself and $\mathcal{B}(\mathcal{a},G)$ has a faithful G-invariant state g such that $g(AU) = 0$ for each A in $\mathcal{a}$ and U in $G\setminus\{I\}$, then Ψ extends to a *-automorphism of $\mathcal{B}(\mathcal{a},G)$ if and only if $\Psi(U)$ is a unitary operator for each U in G.

Using a result of E. Størmer ([49]) we see (Corollary 3.18) that for each isometry Ψ from $\mathcal{B}(\mathcal{a},G)$ onto itself that leaves fixed each operator in $\mathcal{a}$, either $\Psi(U)U^* \in \mathcal{a}$ for each U in G or $U\Psi(U) \in \mathcal{a}$ for each U in G. (A related result was obtained in [29] for a semigroup G generated by a single unitary operator arising from a measure preserving transformaiton).

Next, we utilize what we know about homomorphisms of $\mathcal{g}$ to show that some properties of $\mathcal{g}$, related to the ordering on G (as discussed in Chapter II), are preserved by a homomorphism. We prove (Corollary 3.24) that if the algebra $\mathcal{g}_1$, generated by $\mathcal{a}_1$ and G_1 is isomorphic to the algebra $\mathcal{g}_2$, generated by $\mathcal{a}_2$ and G_2, and if the isomorphism Ψ maps $\mathcal{a}_1$ onto $\mathcal{a}_2$ then G_1 is generated by a single unitary operator if and only if G_2 is generated by a single unitary operator. We also show (Corollary 3.22) that if Ψ can be extended to a *-isomorphism of $\mathcal{B}(\mathcal{a}_1,G_1)$ onto $\mathcal{B}(\mathcal{a}_2,G_2)$ and if G_1 and G_2 are unitary semigroups arising from semigroups of measure preserving transformations of a finite measure space, then G_1 is amenable if and only if G_2 is amenable.

Chapter IV deals with the group of *-automorphisms of $\mathcal{B}(\mathcal{a},G)$ leaving fixed each operator in $\mathcal{a}$ (to be denoted by $\mathrm{Aut}(\mathcal{B},\mathcal{a})$) and the skew adjoint (not necessarily bounded) derivations on $\mathcal{g} + \mathcal{g}^*$ vanishing on $\mathcal{a}$ (denoted by $D(\mathcal{g} + \mathcal{g}^*,\mathcal{a})$).
In this analysis we use the theory of spectrum and spectral spaces associated with an action of a group on a C*-algebra introduced by Arveson in [5].

We define a map $E: D(\mathcal{g} + \mathcal{g}^*,\mathcal{a}) \to \mathrm{Aut}(\mathcal{B},\mathcal{a})$ such that $E(\delta)(T) = (\exp(i\delta))(T)$ for T in $\mathcal{g} + \mathcal{g}^*$.

For a *-automorphism Ψ in $\mathrm{Aut}(\mathcal{B},\mathcal{A})$ we show the equivalence of the following conditions (see Lemma 4.15 and Proposition 4.15):

(1) Ψ is implemented by a unitary operator in $\mathcal{A}$.

(2) Ψ is $\mathcal{E}(\delta)$ for some bounded derivation δ in $D(\mathcal{G} + \mathcal{G}^*,\mathcal{A})$.

(3) Ψ lies in the connected component of the identity in Aut $(\mathcal{B},\mathcal{A})$ (the topology on $\mathrm{Aut}(\mathcal{B},\mathcal{A})$ being the norm topology).

For a derivation δ in $D(\mathcal{G} + \mathcal{G}^*,\mathcal{A})$ we prove the equivalence of the following conditions (see Lemma 4.12 and Theorem 4.20):

(1) δ is bounded.

(2) Then is an operator D in $\mathcal{A}$ such that

$$\delta(T)=DT-TD \qquad \text{for } T \text{ in } \mathcal{G}+\mathcal{G}^*.$$

(3) $\mathrm{Sup}\ \{||\delta(U)|| : U \in G\} < \infty$.

CHAPTER I: INTRODUCING THE MODEL AND SOME EXAMPLES

In this chapter we describe the structure basic to the remaining work. The irreducible triangular algebras to be studied are introduced and some examples are given

We will deal with the action of a semigroup of unitary operators on a maximal abelian von Neumann algebra.

For this we define an ordered (unitary) semigroup to be a semigroup G such that: (1) $G \cup G^{-1}$ is a group, to be denoted by $\bar{G}$.

(2) $G \cap G^{-1} = \{I\}$ where I is the unit element.

(3) For any $W \in \bar{G}$, $WGW^{-1} = G$.

When such G is given it defines an ordering on $\bar{G}$. For U and V in $\bar{G}$ we write $U > V$ if $UV^{-1} \in G$. With U, V and W in $\bar{G}$, we have:

(i) $U > V$ implies $WU > WV$ and $UW > VW$.

(ii) $U > V$ if and only if V^*U is in G (because $V^*U=U^*(UV^*)U$).

Henceforth X will denote a locally compact Hausdorff space and m a σ-finite regular Borel measure on X. Let H be the Hilbert space $L^2(X,m)$ and $B(H)$ be the algebra of all bounded linear operators acting on H.

For any function f in $L^\infty(X,m)$ define the operator L_f in $B(H)$ by $L_f g=fg$ (multiplication by f). The algebra $\mathcal{a} = \{L_f : f \text{ is in } L^\infty(X,m)\}$ is a maximal abelian algebra in $B(H)$. Any unitary operator U that satisfies: $U^* \mathcal{a} U=\mathcal{a}$ is said to <u>act</u> on $\mathcal{a}$, the <u>action</u> being $A \to U^*AU$.

We say that U acts freely on $\mathcal{a}$ if for any non zero projection Q in $\mathcal{a}$, there is a non zero projection E in $\mathcal{a}$ such that $E \leq Q$ and $EU^*EU=0$.

We say that the semigroup G acts freely on $\mathcal{a}$ when each U in G, other then I, acts freely.

Lemma 1.1.: Let W be a unitary operator in $B(H)$ acting on $\mathcal{a}$, then there exists a unique projection F in $\mathcal{a}$ such that:

(1) For any non zero projection Q in $\mathcal{a}$, $Q \leq F$, $QW^*QW \neq 0$ and F is maximal with this property.

(2) F is a maximal projection in $\mathcal{a}$ such that $FW=WF \in \mathcal{a}$. Also, for each subprojection E of F in $\mathcal{a}$ $EW=WE$.

(3) For each non zero projection Q in $\mathcal{a}$, $Q \leq I-F$, there is a non zero subprojection E of Q in $\mathcal{a}$ such that $EW^*EW=0$. The projection F is minimal with this property.

Proof: If W acts freely on $\mathcal{a}$ then $F=0$ clearly satisfies (1) and (3). It also satisfies (2) because if there is a non zero projection F in $\mathcal{a}$ such that $FW=WF \in \mathcal{a}$, then there is a non zero subprojection E of F such that $EW^*EW=0$ (by the free action).
But $EW=EFW=E(FW)=FWE=WFE=WE$ hence $0=EW^*EW=EW^*WE=E$. We assumed that E is a non zero projection and this contradiction proves (2).

If W does not act freely there is a non zero projection F_1 in $\mathcal{a}$ such that for any non zero subprojection Q of F_1 in $\mathcal{a}$ $QW^*QW \neq 0$.
Let $\{F_a\}_{a \in \Lambda}$ be a maximal orthogonal family of projections with this property and set $F=\sum_{a \in \Lambda} F_a$. If Q is a non zero subprojection of F in $\mathcal{a}$ then $QF_a \neq 0$ for some a and $QF_aW^*QF_aW \leq QW^*QW$, so that $QW^*QW \neq 0$. If F_0 has the property that for each non zero subprojection Q of F_0 in $\mathcal{a}$, $QW^*QW \neq 0$, then each non zero subprojection of F_0 has this property.
Now F_0-F_0F is a subprojection of F_0 orthogonal to each projection $F_a (a \in \Lambda)$. By maximality of this family, $F_0-F_0F=0$. Thus $F_0 \leq F$, and F satisfies (1). We see, too, that this F satisfies (3), for, again, by maximality of $\{F_a\}_{a \in \Lambda}$, each non zero subprojection Q of $I-F$ in $\mathcal{a}$ has a non zero subprojection E in $\mathcal{a}$ such that $EW^*EW=0$.
Moreover, if F' is a proper subprojection of F in $\mathcal{a}$ then $F-F'$ is a non zero subprojection Q of F (and of $I-F'$) so that each non zero subprojection of E of Q in $\mathcal{a}$ satisfies: $EW^*EW \neq 0$ (from the properties

of F). Thus F' does not satisfy the property of (3), and F is minimal with this property.

To prove that F satisfies (2), we show, first, that for any non zero subprojection Q of F in $\mathcal{A}$, there is a non zero subprojection E of Q in $\mathcal{A}$ such that $EW=WE\in\mathcal{A}$. Let E be QW^*QW. If E' is a non zero subprojection of E in $\mathcal{A}$ and $W^*E'W \nleq E'$, then $0\neq W^*E'W(I-E')$ $(=E_1) \leq I-E'$ and $0\neq WE_1W^*$ $(=E_2) \leq E'$. But then $E_2W^*E_2W=E_2E_1=0$, contradicting the property of (1) for F (as $0\neq E_2 \leq E \leq Q \leq F$). Thus $W^*E'W \leq E'$. If $W^*E'W < E'$ then $E' < WE'W^*$. Let E_0 be $E'-W^*E'W$ so that $E_0\neq 0$ and $0\neq WE_0W^*=WE'W^*-E' \leq WE'W^* \leq WEW^*=WQW^*Q \leq Q$. Hence $0\neq WE_0W^* \leq Q-E'$, and $W^*E_0WE_0=W^*(E_0WE_0W^*)W \leq W^*E'(Q-E')W=0$. But this contradicts the property of (1) for F. Thus $W^*E'W=E'$. It follows that WE restricted to E(H) commutes with $\mathcal{A}E$. But $\mathcal{A}E$ is maximal abelian on E(H), so the $WE \in \mathcal{A}E$. Hence $WE\in\mathcal{A}$ which establishes our claim.

If $\{E_a\}_{a\,\Lambda}$ is a maximal orthogonal family of subprojections of F in $\mathcal{A}$ commuting with W and such that $WE_a\in\mathcal{A}$, then $F = \sum E_a$ (for otherwise, by what we have just established, $F-\sum E_a$ contains a non zero subprojection E_0 in $\mathcal{A}$ such that $E_0W=WE_0\in\mathcal{A}$ contradicting the maximality of $\{E_a\}_{a\in\Lambda}$). Thus $FW=WF\in\mathcal{A}$.

Finally, we show that F, as constructed, is maximal with the property that $FW=WF\in\mathcal{A}$.

Suppose that $F'\in\mathcal{A}$ and $F'W=WF'\in\mathcal{A}$. As noted above, $QW=WQ\in\mathcal{A}$ for each subprojection Q of F' in $\mathcal{A}$. In particular, this holds for each non zero subprojection Q of F'-FF'. Thus $QW^*QW=Q^2=Q\neq 0$. By maximality of $\{F_a\}_{a\in\Lambda}$, F'-FF' must be 0. It follows that $F' \leq F$, and (2) is established for F. ###

From now on G will always be a semigroup of unitary operators in B(H) and $\mathcal{G}$ (or $\mathcal{A}$ (G)) will be the algebra (not necessarily closed or self adjoint) generated by $\mathcal{A}$ and G.

We will assume that $G \cap \mathcal{A} = \{I\}$.

Note that if G is an ordered semigroup and U,V are in G and are different from I (the unit element) then also $UV \neq I$.

The following proposition extends a result of R.V.Kadison and I.M.Singer in [33, Theorem 2.2.1].

Proposition 1.2.: If G is an ordered semigroup of unitary operators each of which, other the I, acts freely on $\mathcal{a}$ then the algebra $\mathcal{g}$ is triangular.

Proof: Let T be a self adjoint operator in $\mathcal{g}$.

Then $T=A_0+A_1U_1+\ldots+A_nU_n=T^*=A_0^*+U_1^*A_1^*+ \ldots +U_n^*A_n^*$

$$=A_0^*+B_1U_1^*+ \ldots +B_nU_n^*,$$

where $A_o, \ldots, A_n, B_1, \ldots, B_n \in \mathcal{a}$, $U_j \neq U_k$ $(j \neq k)$ and $I < U_1 < U_2 < \ldots < U_n$; and

$$0=B_n+B_{n-1}U_{n-1}^*U_n+ \ldots +B_1U_1^*U_n+(A_0^*-A_0)U_n-A_1U_1U_n-\ldots-A_nU_n^2 . \qquad (*)$$

Note that

$$I < U_{n-1}^*U_n < \ldots < U_1^*U_n < U_n < U_1U_n < \ldots < U_{n-1}U_n < U_n^2.$$

We refer to an equation of the form:

$0=C_0+C_1V_1+C_2V_2+ \ldots +C_mV_m$, where $C_0,\ldots,C_m \in \mathcal{a}$ and $I < V_1 < \ldots < V_m$, $V_j \neq V_k$ $(j \neq k)$, as a dependence relation (for G over $\mathcal{a}$) of length m; so that (*) is such a relation of length at least 1 unless $T=A_0$ $(=A_0^*)$ and $n=0$.

Let $0=A_0+A_1U_1+\ldots+A_nU_n$ be a dependence relation of minimal length and let E be the range projection of A_n. We prove that E is 0 (as a consequence of the free action of U_n on $\mathcal{a}$), whence $A_n=0$ and the only dependence relation is $0 = 0$. It will follow from this that $T=A_0$ and $\mathcal{g}$ is triangular.

Suppose $F \in \mathcal{a}$ and $F \leq E$. Let Q be $U_nFU_n^*$. We show that $Q \leq F$. If this is not the case, then $M=Q-QF$ is a non zero projection smaller than Q and $0 < U_n^*MU_n(=N) \leq F \leq E$. Hence $NA_n \neq 0$ and, $0=NA_0+NA_1U_1+\ldots+NA_nU_n$, is another dependence relation (for G over $\mathcal{a}$). Now $NA_0 \neq 0$, by minimality of n (for otherwise, $0=NA_1+NA_2U_2U_1^*, +\ldots+NA_nU_nU_1^*$ is a "smaller" dependence relation).

But $0=N(A_0+A_1U_1+\ldots+A_{n-1}U_{n-1}+A_nU_n)N=NA_0+B_1U_1+\ldots+B_{n-1}U_{n-1}+NA_nMU_n$, and $NA_nM=FNA_nM=MFNA_n=0$ (as $MF=0$). Hence $Q \leq F$; and $Q=QF=UFU_n^*F$.
If $E\neq 0$ and F is chosen as a non zero subprojection of E in $\mathcal{a}$, then $0\neq U_n^*QU_n=FU_n^*FU_n$, which contradicts the free action of U_n on $\mathcal{a}$. Thus $E=0$. ###

In fact we have proved that if $A_0+A_1U_1+\ldots+A_nU_n=0$ where $A_i\in\mathcal{a}$, $U_i \in G\setminus\{I\}$, and $U_i\neq U_j$ when $i\neq j$ then $A_i=0$ for all i. We derive from this the following:

Corollary 1.3.: If G acts freely on and $0=\sum_{k=1}^m B_kV_k+\sum_{j=0}^n A_jU_j$, where $B_k,A_j\in\mathcal{a}$, $I < U_0 < \ldots < U_n$, $U_i \in G$, $U_i\neq U_j$ when $j\neq i$, $V_k^* \in G\setminus\{I\}$, $V_1^* < V_2^* < \ldots < V_m^*$, and $V_k\neq V_i$ when $k\neq i$ then $B_k=A_j=0$ for all k and j.

Proof: From our hypothesis, we have
$0=\sum_{k=1}^m B_kV_kV_m^*+\sum_{j=0}^n A_jU_jV_m^*$, $V_kV_m^*$, $U_jV_m^* \in G$, $V_kV^*_m\neq U_jV_m^*$ for all k and j, $V_kV_m^*\neq V_iV_m^*$ when $k\neq 1$, $U_jV_m^*\neq U_iV_m^*$ when $i\neq j$.
Thus we are in the situation of Proposition 1.2 and can conclude (from its proof) that $B_k=A_j=0$ for all k and j. ###

We refer to the property described in this Corollary as $\mathcal{a}$-independence of $\bar{G}$.

Definition: We say that G acts ergodically on $\mathcal{a}$ (or that G is ergodic) if for each non zero projection P in $\mathcal{a}$, $I=\vee\{U^*PU:U \in G\}$.

Lemma 1.4.: For a (unitary) semigroup acting on $\mathcal{a}$ the following are equivalent:

(1) G is ergodic.

(2) $G^{-1}(=G^*)$ is ergodic.

(3) For every projection $Q \in \mathcal{a}$: $QUQ=UQ$ for all U in G implies $Q=0$ or $Q=I$.

Proof:

(1) implies (2): For a projection P in $\mathcal{a}$ let $Q=\vee\{UPU^* : U \in G\}$ then

$VQV^* \leq Q$ for all V in G. Therefore $I-Q \geq V^*(I-Q)V$ for all V in G, hence $I-Q \geq \bigvee\{V^*(I-Q)V : V \in G\}$. If $Q \neq I$ then, by (1), $I-Q \geq I$ hence $Q=0$ and so $P=0$.

Therefore, for $P \neq 0$, $Q=I$ which proves (2).

(2) implies (3) : We assume $QUQ=UQ$ for all U in G. For all U in G $QUQU^*=UQU^*$ hence $Q \leq UQU^*$, but this, by (2), implies that $Q=0$ or $Q=I$.

(3) implies (1): For a non zero projection P in $\mathcal{a}$ let $Q = \bigvee\{U^*PU : U \in G\}$. then $V^*QV \leq Q$ for all V in G, hence $V^*(I-Q)V=I-V^*QV \geq I-Q$ so $V^*(I-Q)V(I-Q)=I-Q$ and $(I-Q)V(I-Q)=V(I-Q)$. This, by (3), shows that $Q=0$ or $Q=I$ but, since $Q \geq P \neq 0$, $Q=I$. ###

<u>Definition:</u> An algebra $\mathcal{g}$ of operators on a Hilbert space H is called irreducible if $\text{Lat}\mathcal{g} = \{0,I\}$ where $\text{Lat}\mathcal{g} = \{P \in B(H) : PTP=TP$ for every T in $\mathcal{g}\}$.

<u>Corollary 1.5.</u>: Let G be an ordered semigroup acting freely and ergodically on $\mathcal{a}$. Then $\mathcal{g}$ is an irreducible triangular algebra.

<u>Proof</u>: $\mathcal{g}$ is triangular by Proposition 1.2, and is irreducible by (3) of Lemma 1.4. ###

<u>Lemma 1.6.</u>: Let G be an abelian ordered semigroup that acts ergodically on $\mathcal{a}$. Then $\mathcal{g}$ is an irreducible triangular algebra.

<u>Proof</u>: What we have to prove is that G acts freely on $\mathcal{a}$.

If U is in $G \setminus \{I\}$ but does not act freely, then, by (2) of Lemma 1.1, there exists a non zero projection E in $\mathcal{a}$ such that $EU=UE$ and EU is in $\mathcal{a}$.

Since U is not in $\mathcal{a}$, and $\mathcal{a}$ is maximal abelian, there is a projection F in $\mathcal{a}$ such that $U^*FU \neq F$ and we can assume $(I-F)U^*FU \neq 0$ (if $U^*FU < F$ take $FU^*(I-F)U$ instead). Let Q be $(I-F)U^*FU$ then $QU^*QU=0$. For V in G let Q_V be V^*QV. Then

$$Q_V U^* Q_V U = V^*QVU^*V^*QVU = V^*(QU^*QU)V = 0.$$

Since $I=\bigvee\{Q_V : V \in G\}$, there exists V in G such that $Q_V E \neq 0$. Now $U^* Q_V U Q_V = Q_V U^* Q_V U = 0$; so that

$$Q_V UEQ_V EU = U(U^* Q_V UEQ_V E)U = U(U^* Q_V UQ_V E)U = 0.$$

Using the fact that $EU=UE$ and EU is in $\mathcal{a}$ we see that $Q_V EU=0$ thus $Q_V E=0$, and this contradiction completes the proof. ###

Let $\mathcal{a}$ be as before and let G be the semigroup generated by a single unitary operator U acting on $\mathcal{a}$. Then G acts ergodically if and only if, for every projection P in $\mathcal{a}$ such that $PU^nP=U^nP$, for all n, $P=0$ or $P=I$.

For ergodicity it is equivalent to assume that $PUP=UP$ implies $P=0$ or $P=I$, since $PUP=UP$ implies

$$PU^nP=PU^{n-1}UP=PU^{n-1}PUP= \dots =PUPUP\dots UP=UPUP\dots UP=u^2PUP\dots P=U^3P\dots P=\dots=U^nP.$$

We can, therefore, conclude (see [33, Theorem 2.2.1]):

Lemma 1.7.: If U is a unitary operator acting ergodically on an infinite dimensional maximal abelian algebra $\mathcal{a}$, then $\mathcal{g}$ (the algebra generated by U and $\mathcal{a}$) is an irreducible triangular algebra.

Proof: As noted above, the semigroup G, generated by U, acts ergodically on $\mathcal{a}$. We can, therefore, apply Lemma 1.6 provided G is an ordered semigroup.

For this we have to show that $G \cap G^* = \{I\}$.

If this is not so, there are positive integers m and p such that $U^m=U^{-p}$; so that $U^{m+p}=I$.

Let n be the least positive integer such that $U^n=I$. We note the (easy) fact that a finite group $\{U^j : j=1,2,\dots,n\}$ cannot act ergodically on the infinite-dimensional maximal abelian algebra $\mathcal{a}$.

If we assume that U does act ergodically on $\mathcal{a}$ and E is a non zero projection in $\mathcal{a}$, then

$$I=\bigvee_{j=1}^{n} U^{*j}EU^j \leq U^*EU+ \dots +U^{*n}EU^n.$$

If E is a minimal projection in $\mathcal{a}$, then I is the union of n minimal projections - contradicting the fact that $\mathcal{a}$ is infinite dimensional. Since $\mathcal{a}$ has no minimal projections, if x_0 is a unit vector in H, there is a projection E in $\mathcal{a}$ such that $<Ex_j,x_j> < \frac{1}{n}$ $(j=1,\dots,n)$ where $x_j=U^j x_0$. But then

$$1= < x_0,x_0> \leq \sum_{j=1}^{n} <Ex_j,x_j> <1$$

- a contradiction ###

Examples of irreducible triangular algebras

Example 1.8.: (See [33, Example 2.2.2]) Let $X = \mathbb{T}$ (the unit circle) and m be the Haar measure on $\mathbb{T}$.

Let U be the unitary operator on $H=L^2(X,m)$ induced by an irrational translation (i.e. $Uf(e^{2\pi it})=f(e^{2\pi i(t-s)})$ where s is a real irrational number). For f in $L^\infty(X,m)$, L_f is in $\mathcal{A}$ and $U^*L_fU=L_{f\circ\tau^{-1}}$ where τ is the transformation $\tau(e^{2\pi it})=e^{2\pi i(t-s)}$ on X, hence $U^*\mathcal{A}\ U=\mathcal{A}$.

To prove that U acts ergodically notice that each projection P in $\mathcal{A}$ can be written as L_f where f is the characteristic function of a set E in X.

The condition $PUP=UP$ (or $U^*PUP=P$) yields $fg=f$ where $g(=f\circ\tau^{-1}=Uf)$ is the characteristic function of $\tau(E)$. This, using the fact that $m(E)=m(\tau(E)) < \infty$, implies that $f=g$. Hence $Uf=f$.

The function f has an expansion, $f=\sum_{n=-\infty}^{\infty}a_ne_n$, where $\{e_n\}_{n=-\infty}^{\infty}$ is the orthonormal basis of $L^2(X,m)$, $e_n(e^{it})=e^{nit}$. But $Uf=\sum_{n=-\infty}^{\infty}c^na_ne_n=f=\sum_{n=-\infty}^{\infty}a_ne_n$, where $c=e^{-2\pi is}$; so that, for each integer n, $c^na_n=a_n$. By choice of s, $c^n=1$ only when $n=0$. Thus $f=a_0$, P is 0 or I, and U acts ergodically on $\mathcal{A}$. By Lemma 1.7 the algebra generated by U and $\mathcal{A}$ is an irreducible triangular algebra.

Example 1.9.: Let X be $\mathbb{R}$ (the set of all real numbers) and m be Lebesgue measure on $\mathbb{R}$.

Fix a negative irrational number u and define the set T in $\mathbb{R}$: $T=\{au-r : a > 0,\ a\in\mathbb{Q},\ r\geq 0,\ r\in\mathbb{Q}\} \cup \{bu+r: r,b\in\mathbb{Q},\ b\geq 0,\ r\geq 0\}$.

Let G be the semigroup of translations by t in T, i.e. U in G is of the form $Uf(s)=f(s-t)$ for some t in T, where $f\in L^2(X,m)$.

We claim that G satisfies the conditions of Lemma 1.6:

1) G is clearly a semigroup.

2) G is an ordered semigroup because $T\cap(-T)=\{0\}$ and $T\cup(-T)=u\mathbb{Q}+\mathbb{Q}$.

3) $U^*\mathcal{A}\ U=\mathcal{A}$, for if U in G is the operator corresponding to t in T

and f is in $L^\infty(X,m)$, then $U^*L_fU=L_{f_{-t}}$ where $f_t(s)=f(s-t)$.

4) The action of G is ergodic. This fact will be proved for a more general case in Proposition 1.10 below, and is omitted here.

This shows that the algebra $\mathcal{G}$ generated by G and $\mathcal{a}$ is irreducible and triangular.

The following Proposition generalizes Examples 1.8 and 1.9.

Proposition 1.10.: Let N be a locally compact group with a left invariant Haar measure m. Let T be an ordered semigroup of N which is dense in N. For each x in N define the unitary operator U_x on $L^2(N,m)$ by a left translation: $(U_xf)(y)=f(x^{-1}y)$. Denote by G the semigroup of the unitary operators U_x for x in T. Then the algebra generated by G and the multiplication algebra $\mathcal{a}$ is irreducible triangular.

Proof : We will show that G satisfies the conditions of Corollary 1.5.

Let A be the operator in $\mathcal{a}$ corresponding to multiplication by a function g(in $L^\infty(N,m)$). Then

$(U^*_xAU_xf)(y)=(AU_xf)(xy)=g(xy)(U_xf)(xy)=g(xy)f(y)=g_x(y)f(y)$ where $g_x(y)=g(xy)$.

Thus $U^*_xAU_x=A_x$, where A_x corresponds to multiplication by g_x; and $U^*_x\, \mathcal{a}\, U_x=\mathcal{a}$ for each x in N.

We show, next, that each $U_x(x\neq e)$ acts freely on $\mathcal{a}$.

Suppose the contrary. Then, from (2) of Lemma 1.1, there is a non zero projection E in $\mathcal{a}$ such that $FU_x=U_xF \in \mathcal{a}$ for each subprojection F of E in $\mathcal{a}$. Let Y be a Borel set in N corresponding to E (i.e. E is multiplication by the characteristic function of Y).

Since $E\neq 0$, $m(Y) > 0$; and there is a compact subset K of Y such that $m(K) > 0$ (by regularity of m). As $x\neq e$, we can choose a symmetric open neighbourhood V of e such that $x \notin V^2$. Then $Vy \cap x^{-1}Vy=\emptyset$ for each y in N. Now $\{Vy : y \in K\}$ is an open covering of K from which we can select a finite subcovering.

Thus $m(Vy \cap K) > 0$ for some y in K. Let F be the projection corresponding to multiplication by the characteristic function of $Vy \cap K$ $(=Z)$; so that F is a non zero subprojection of E in $\mathcal{a}$, and $U_x^*FU_x=F$. But $U_x^*FU_x$ is the projection corresponding to multiplication by the characteristic function of $x^{-1}Z$ and

$$Z \cap x^{-1}Z \subseteq Vy \cap x^{-1}Vy=\emptyset. \quad \text{Thus} \quad FU_x^*FU_x=F^2=F=0$$

a contradiction. Hence U_x acts freely on $\mathcal{a}$.

To prove the ergodicity of G we, first, show that, for a non zero projection $P \in \mathcal{a}$, $I = \bigvee\{U_x^*PU_x \ ; \ x \in N\}$. For this, let Q be $\bigvee\{U_x^*PU_x; \ x \in N\}$ and let Y be a set of positive measure such that Q is the projection corresponding to multiplication by the characteristic function of Y.

Define a measure on the Borel sets of N by

$$m_1(Z)=m(Z \cap Y).$$

Since $m_1(xZ)=m(xZ \cap Y)=m(Z \cap x^{-1}Y)=m(Z \cap Y)+m(Z \cap (x^{-1}Y \setminus Y))$

$$-m(Z \cap (Y \setminus x^{-1}Y))=m(Z \cap Y)=m_1(Z) \quad (\text{as} \quad m(Y \setminus x^{-1}Y)=m(x^{-1}Y \setminus Y)=0,$$

it follows that $m_1(xZ)=m_1(Z)$.

Thus m_1 is a left Haar measure provided $m_1(U) > 0$ for every non empty open set U in N. This condition is equivalent to the assertion that m_1 is not identically zero (see [27, p. 252]). Indeed, if $m_1(U)=0$ for some non empty open set U, and if C is any compact set then the class $\{xU : x \in C\}$ is an open covering of C. Since C is compact, there exists a finite subset $\{x_1,\ldots,x_n\}$ of C such that $C \subseteq \bigcup_{i=1}^n x_iU$, and the left invariance of m_1 implies $m_1(C) \leq \sum_{i=1}^n m_1(x_iU)=n\, m_1(U)$.

It follows that $m(C)=0$ for every compact set C contained in Y, and this, by regularity, implies that $m(Y)=0$. This contradicts our assumption and completes the proof that m_1 is a left Haar measure on N.

Therefore $m_1=cm$, for some constant c, and since m_1 vanishes on the complement of Y, so does m. This proves that $Q=I$.

We will now show that for a non zero projection P in $\mathcal{a}$,

$I = \bigvee\{U_x^*PU_x \ : \ x \in G\}$.

For this, let Q be $I-\bigvee\{U_x^*PU_x \ : \ x \in G\}$.

Choose any f in $L^2(N,m)$ such that $Qf=f$. It is known that the map $x \to U_x f$, from N to $L^2(N,m)$, is continuous.

Thus the map $x \to ||U_x^* P U_x f||^2 = \langle U_x^* P U_x f, f\rangle = \langle PU_x f, U_x f\rangle$ is continuous. Since $P_x Q=0$ for each x in G, $U_x^* P U_x f=0$. As G is dense in N, the continuity implies that $U_x^* P U_x f=0$ for each x in N. This is possible only if $f=0$ (since $I = \vee\{U_x^* P U_x \; ; \; x \in N\}$).

Consequently $Q=0$ and the action is ergodic.

The conditions of Corollary 1.5 are satisfied; and by that corollary the algebra generated by G and the multiplication algebra is irreducible triangular. ###

The last proposition enables us to present an example of an irreducible triangular algebra where the semigroup G is non abelian.

Example 1.11.: Let N be the group of 2x2 matrices of the form: $\begin{pmatrix} a & b \\ 0 & 1 \end{pmatrix}$ where a and b are real numbers and $a > 0$. For x in N we denote by $a(x)$ and $b(x)$ the numbers such that $x = \begin{pmatrix} a(x) & b(x) \\ 0 & 1 \end{pmatrix}$.

As $\begin{pmatrix} a & b \\ 0 & 1 \end{pmatrix} \begin{pmatrix} c & d \\ 0 & 1 \end{pmatrix} = \begin{pmatrix} ac & ad+b \\ 0 & 1 \end{pmatrix} \in N$

and $\begin{pmatrix} a & b \\ 0 & 1 \end{pmatrix}^{-1} = \begin{pmatrix} a^{-1} & -ba^{-1} \\ 0 & 1 \end{pmatrix} \in N$;

N is a group. We denote the unit element by e. We can identify N, as a set, with a subset of $\mathbb{R}^2$ and the topology on N is induced from $\mathbb{R}^2$ i.e. $x_n \to x$ if and only if $a(x_n) \to a(x)$ and $b(x_n) \to b(x)$. With this topology, N is a locally compact group.

Let M_1 be any dense, ordered, subsemigroup of the additive group of all the real numbers (e.g. the semigroup T of Example 1.9).

Let M denote the multiplicative semigroup $\exp(M_1)$, hence $M \cap M^{-1} = \{1\}$.

Let G be the set of all elements x in N for which either $a(x) \in M\setminus\{1\}$ or $a(x)=1$ and $b(x) \geq 0$.

It is easy to verify that G is a semigroup. In fact, an ordered semigroup. This can be seen as follows:

1) If x is in $G \cap G^{-1}$ then $a(x) \in M \cap M^{-1}=\{1\}$ and therefore $b(x) \geq 0$. But $b(x^{-1})=-b(x)a(x)^{-1} \leq 0$ hence x^{-1} is in G only if $b(x)=0$, and then $x=e$.

2) $G \cup G^{-1}$ is the set of the elements x in N such that $a(x) \in M \cup M^{-1}$ and $b(x)$ is a real number. Since $M \cup M^{-1}$ is a multiplicative group, $G \cup G^{-1}$ is a group.

3) For every x in $G \cup G^{-1}$ and y in G,

$$x = \begin{pmatrix} a & b \\ 0 & 1 \end{pmatrix}, \quad x^{-1} = \begin{pmatrix} a^{-1} & -ba^{-1} \\ 0 & 1 \end{pmatrix}, \quad y = \begin{pmatrix} c & d \\ 0 & 1 \end{pmatrix}$$

and $x^{-1}yx = \begin{pmatrix} c & (cb+d-b)a^{-1} \\ 0 & 1 \end{pmatrix}$

if $c \neq 1$ then $x^{-1}yx$ is in G, and if $c=1$ then $d \geq 0$ and $(cb+d-b)a^{-1}=da^{-1} \geq 0$. Hence $x^{-1}yx$ is in G in any case.

This shows that G is an ordered semigroup and clearly G is dense in N, therefore, by Proposition 1.10, the algebra generated by the semigroup of unitary operators associated with G, and the multiplication algebra on $L^2(N,m)$ is an irreducible triangular algebra.

<u>Example 1.12.</u> : Let U be a unitary operator on H acting ergodically on $\mathcal{a}$. Let H_2 be $H \oplus H$. Each operator in $B(H_2)$ can be written as a 2x2 matrix with entries in $B(H)$.

Let V in $B(H_2)$ be the operator $\begin{pmatrix} 0 & U \\ I & 0 \end{pmatrix}$ and then:

$$V^2 = \begin{pmatrix} U & 0 \\ 0 & U \end{pmatrix} \qquad V^3 = \begin{pmatrix} 0 & U^2 \\ U & 0 \end{pmatrix} \qquad \text{etc.}$$

$\mathcal{a}_2 = \{ \begin{pmatrix} A_1 & 0 \\ 0 & A_2 \end{pmatrix} : A_i$ for $i=1,2\}$ is a maximal abelian algebra.

Since $V^* \begin{pmatrix} A_1 & 0 \\ 0 & A_2 \end{pmatrix} V = \begin{pmatrix} A_2 & 0 \\ 0 & U^*A_1U \end{pmatrix}$ is in $\mathcal{a}_2$, $V^* \mathcal{a}_2 V = \mathcal{a}_2$.

Each projection P in $\mathcal{a}_2$ has the form $\begin{pmatrix} P_1 & 0 \\ 0 & P_2 \end{pmatrix}$,

and $PVP=VP$ when $\begin{pmatrix} 0 & UP_2 \\ P_1 & 0 \end{pmatrix} = \begin{pmatrix} 0 & P_1UP_2 \\ P_2P_1 & 0 \end{pmatrix}$ (P_i are projections in $\mathcal{a}$).

Thus $PVP=VP$ if and only if $P_1 \leq P_2$ and $P_1UP_2=UP_2$ (hence also $P_2UP_2=UP_2$). Since U acts ergodically we get $P_2=0$ or $P_2=I$.

If $P_2=0$ then, since $P_1 \leq P_2$, $P_1=0$ hence $P=0$.

If $P_2=I$ then, since $P_1 U P_2 = UP_2$, $P_1=1$; and $P=I$.

Therefore V acts ergodically on a_2 and the algebra generated by a_2 and V is triangular and irreducible. This example is different from Example 1.8 because V^2 is not ergodic, while in Example 1.8 any power of the generator is also ergodic.

Example 1.13.: Let U be a unitary operator on H acting ergodically on a. Define $K = \sum_{i=1} \oplus H_i$ and $H^{(n)} = \sum_{i=1}^{2n-1} \oplus H_i$ where $H=H_i$ $(H^{(k+1)}=H^{(k)} \oplus H^{(k)})$. Let E_i and F_i be the projections from K onto H_i and $H^{(i)}$ respectively, and let $\bar{a}$ the maximal abelian algebra of all operators A in $B(K)$ such that for any positive integer n $E_n A = AE_n$ and $E_n AE_n$ is in a.
We define algebras g_n by induction. Let g_1 be g (the algebra generated by U and a on H).
Let g_2 on $H^{(2)}$ $(=H\oplus H)$ be generated by $a \oplus a$ and the operator represented by the matrix $\begin{pmatrix} 0 & U \\ I & 0 \end{pmatrix}$. Then

g_2 is irreducible triangular (Example 1.12) and $g_1 = E_1 g_2 E_1 = E_2 g_2 E_2$.
Assume that g_k is defined and let g_{k+1} on $H^{(k+1)}$ be generated by the algebra $a_k \oplus a_k$ $(a_k = g_k \cap g_k^*)$, and the operator U_{k+1} represented by the matrix $\begin{pmatrix} 0 & U_k \\ I & 0 \end{pmatrix}$ where U_k is the generator of g_k.
Define $\bar{g}$ (on $B(K)$) to be the algebra generated by $\cup_{n=1}^{\infty} g_n$ and $\bar{a}$, where we identify g_n with $F_n g_n F_n \subseteq B(K)$. With T_n in g_n and A in $\bar{a}$, $T_n A = T_n F_n A = T_n F_n AF_n \in g_n$. Similarly $AT_n \in g_n$; so that each operator in $\bar{g}$ has the form $A+T_n$ with A in $\bar{a}$ and T_n in g_n for some n. If $A+T_n$ is self adjoint, so is $F_m(A+T_n)F_m$. Now $F_m(A+T_n)F_m \in g_m$ so that $F_m(A+T_n)F_m$ is in $a_m (\subseteq \bar{a})$, for all m.
Since $F_m \uparrow I$, $A+T_n \in \bar{a}$ and $\bar{g}$ is triangular.

If P is a non zero projection in $\bar{a}$ invariant under $\bar{g}$, then, since $F_n \uparrow I$, $F_n P$ is a non zero projection for some n (and, hence, for all larger n). Now $F_n P$ is in g_n and invariant under g_n. Since g_n is irreducible,

$F_nP=F_n$; and $F_n \leq P$. Now $F_nP \leq F_mP$ if $m \geq n$; whence $F_m \leq P$ if $m \geq n$, and $P=I$. Thus $\bar{\mathcal{g}}$ is irreducible.

Therefore $\bar{\mathcal{g}}$ is an irreducible triangular algebra but is not described by Corollary 1.5 because no operator in $\bar{\mathcal{g}}$ is acting freely on $\mathcal{a}$.

CHAPTER II: THE STRUCTURE OF THE IRREDUCIBLE TRIANGULAR ALGEBRAS

In this chapter we will study the structure and properties of the algebra $\mathcal{G}$ generated by a semigroup G of unitary operators and a maximal abelian algebra $\mathcal{a}$, where and G satisfy the conditions of Corollary 1.5. By that corollary the algebra is irreducible and triangular.

We will deal also with the algebra $\mathcal{G}+\mathcal{G}^*$ which is the self adjoint algebra generated by $\mathcal{G}$, and the algebra $\mathcal{B}(\mathcal{a}, G)$ which is the closure of $\mathcal{G}+\mathcal{G}^*$ in the norm topology. The algebra $\mathcal{B}(\mathcal{a}, G)$ is a C*-algebra and, when it does not cause any confusion, we will denote it by $\mathcal{B}(\mathcal{G})$ or just $\mathcal{B}$.

The map Φ.

The following results, from Lemma 2.1 to Lemma 2.9, are Arveson's and appeared in [3].

Recall (Corollary 1.3) that if $A_0+A_1U_1+\ldots+A_nU_n=0$ for $U_i \in \bar{G}\setminus\{I\}$, $A_i \in \mathcal{a}$, and $U_i \neq U_j$ for $i \neq j$, then $A_i=0$ for all i.

This enables us to define a map Φ from $\mathcal{G}+\mathcal{G}^*$ onto $\mathcal{a}$ by:

$\Phi(A_0+A_1U_1+\ldots+A_nU_n)=A_0$ $\quad A_i \in \mathcal{a}$, $U_i \in \bar{G}\setminus\{I\}$. Clearly Φ is linear and onto .

Lemma 2.1:

(1) $\Phi \circ \Phi=\Phi$ and $\Phi(I)=I$.

(2) $\Phi(AT)=A\Phi(T)$, $\Phi(TA)=\Phi(T)A$ for $A \in \mathcal{a}$, $T \in \mathcal{G}+\mathcal{G}^*$.

(3) $\Phi(T^*)=\Phi(T)^*$ for $T \in \mathcal{G}+\mathcal{G}^*$.

(4) $\Phi(U)=0$ for U in $\bar{G}\setminus\{I\}$.

(5) $0 \leq \Phi(T)^*\Phi(T) \leq \Phi(T^*T)$ for $T \in \mathcal{G}+\mathcal{G}^*$.

Proof: (1)-(4) are clear from the definition. To prove (5), let T be $A_0+A_1U_1+\ldots+A_nU_n$ with $A_i \in \mathcal{a}$, $U_i \in \bar{G}$. $T^*=A_0^*+U_1^*A_1^*+\ldots+U_n^*A_n^*$ and $T^*T=A_0^*A_0+U_1^*A_1^*A_1U_1+\ldots+U_n^*A_n^*A_nU_n+S$ where $\Phi(S)=0$. Therefore:

$\Phi(T^*T)=A_0^*A_0+U_1^*A_1^*A_1U_1+\ldots+U_n^*A_n^*A_nU_n \geq A_0^*A_0=\Phi(T)^*\Phi(T) \geq 0.$ ###

Lemma 2.2. : For any non zero projection P in $\mathcal{a}$, there exists a state ρ of $\mathcal{B}(\mathcal{a},G)$ such that $\rho(P)=1$ and $\rho(\mathcal{a}U)=0$ for each $U\in\bar{G}\setminus\{I\}$.

Proof: Fix a non zero projeciton P in $\mathcal{a}$. Let Γ be the collection of all finite subsets of G. For any $F\in\Gamma$, let K_F be the set of all states ρ of $\mathcal{B}(\mathcal{a},G)$ such that $\rho(P)=1$ and $\rho(\mathcal{a}U)=0$ for every U in F. Each K_F is a weak* closed, hence compact, subset of the state space of $\mathcal{B}(\mathcal{a},G)$ and $F_1\subseteq F_2$ implies $K_{F_1}\supseteq K_{F_2}$.

By free action, there exists, for each F in Γ, a non zero projection Q_F in $\mathcal{a}$ such that $Q_F\leq P$ and $U^*Q_FUQ_F=0$ for U in F. Choose any unit vector f in the range of Q_F and let: $\rho_F(T)=\langle Tf,f\rangle$ for $T\in\mathcal{B}(\mathcal{a},G)$.

Clearly $\rho_F(P)=1$ and one has $\rho_F(Q_FTQ_F)=\rho_F(T)$ $T\in\mathcal{B}(\mathcal{a},G)$. If $A\in\mathcal{a}$ and $U\in F$: $\rho_F(AU)=\rho_F(Q_FAUQ_F)=\rho_F(Q_FUQ_FU^*AU)=0$.

Thus $\rho_F\in K_F$ and this shows that the K_F's are non empty and have the finite intersection property. Hence $\cap\{K_F : F\in\Gamma\}\neq\emptyset$, and any state in the intersection has all the required properties. ###

Proposition 2.3.: $||\Phi(T)||\leq||T||$ for $T\in\mathcal{g}+\mathcal{g}^*$.

Proof: Let T be in $\mathcal{g}+\mathcal{g}^*$, and suppose that $||\Phi(T)||>||T||$. Since $\Phi(T)\in\mathcal{a}$, there is a unitary operator $W\in\mathcal{a}$ such that $\Phi(T)W\geq 0$. Now $||\Phi(T)W||=||\Phi(T)||\geq||T||+\varepsilon$, for sufficiently small $\varepsilon>0$. So by spectral theory there exists a non zero projection $P\in\mathcal{a}$ such that $\Phi(T)WP\geq(||T||+\frac{1}{2}\varepsilon)P$. Now choose ρ as in Lemma 2.2 : $\rho(P)=1$, $\rho(\mathcal{a}U)=0$ $U\in\bar{G}\setminus\{I\}$. By the definition of Φ, it follows that $\Phi(S)=\rho\circ\Phi(S)$ for $S\in\mathcal{g}+\mathcal{g}^*$. Hence

$\rho(TWP)=\rho\circ\Phi(TWP)=\rho(\Phi(T)WP)\geq(||T||+\frac{1}{2}\varepsilon)\rho(P)=||T||+\frac{1}{2}\varepsilon.$

But this is impossible, since: $|\rho(TWP)|\leq||TWP||\leq||T||$ and this contradiction completes the proof. ###

We can now extend Φ to $\mathcal{B}(\mathcal{a},G)$ and Lemma 2.1 is still valid for the extension which we denote, again, by Φ.

For the next result we will need the concept of amenable group.

Definition: A (discrete) group N is said to be amenable if there exists a finitely additive probability measure on the field of all subsets of N such that $\mu(xE) = \mu(E)$ for all $X \in N$, $E \subseteq N$.
When such a μ exists, it is called a (left invariant) mean.
Examples of amenable groups are the solvable and the locally finite groups. In particular any commutative group is amenable. An example of a non amenable group is the free group on $n \geq 2$ generators.

Let $C_{00}(N)$ be the set of complex valued functions on N having a finite support, and let $x \to 1_x$ be the left regular representation of N in $l^2(N)$ (i.e. $1_x g(y)=g(x^{-1}y)$, $g \in l^2(N)$).
For any $f \in C_{00}(N)$, we can form the operator $T = \sum_x f(x)1_x$ on $l^2(N)$; f is said to be of positive type if $T \geq 0$.
We state the following lemma:

Lemma 2.4.: The following are equivalent, for any discrete group N:

(1) N is amenable.

(2) For any finite set $F \subseteq N$ and $\varepsilon > 0$, there is a finite subset $E \subseteq N$ such that $\left| \frac{|xE \cap E|}{|E|} - 1 \right| \leq \varepsilon$, for all $x \in F$ ($|\cdot|$ denotes "number of elements in").

(3) If $f \in C_{00}(N)$ is of positive type, then $\sum f(x) \geq 0$.

(4) For any finite subset F of N and $\varepsilon > 0$, there exists a unit vector $g \in l^2(N)$ such that for all $x \in F$ $|<1_x g, g> -1| \leq \varepsilon$.

The equivalence of (1) and (2) was proved by Følner [21], and the rest of the proof can be found in [3, Lemma 2.2].

We shall now construct another C*-algebra $\mathcal{C}$ and a faithful positive map $w:\mathcal{C} \to \mathcal{a}$. We then define a *-homomorphism Π_0 of a dense *-subalgebra of $\mathcal{C}$ into $\mathcal{B}(\mathcal{a},G)$ such that $\Phi \circ \Pi_0 = w$ and we shall analize the extensions of Π_0 to $\mathcal{C}$.

Let us form the space $H \otimes l^2(\bar{G})$. For $A,B \in \mathcal{a}$ and $U,V \in \bar{G}$, $(AU \otimes 1_U)(BV \otimes 1_V)=AUBV \otimes 1_{UV}=AUBU^*UV \otimes 1_{UV}$ and $(AU \otimes 1_U)^*=U^*A^*UU^* \otimes 1_{U^*}$. Thus $C_0=\{\sum_{U\in F} A_U U \otimes 1_U : F \subseteq \bar{G}$ is finite$\}$ forms a $*$-algebra with identity. Let C be the norm closure of C_0 in $B(H \otimes l^2(\bar{G}))$.

Let g_0 in $l^2(\bar{G})$ be defined by $g_0(I)=1$, $g_0(U)=0$ for $U\neq I$.

<u>Lemma 2.5.:</u> There exists a faithful positive linear map $w:C \to \mathcal{a}$ such that

$$\langle w(C)f,g\rangle=\langle C(f \otimes g_0), g \otimes g_0\rangle$$ for any $f,g \in H$. One has $w(\sum_{U\in F} A_U U \otimes 1_U) = A_I$ for any finite subset F of $\bar{G}$.

<u>Proof:</u> For any $C\in C$ define a bilinear form on $H \times H$ by: $[f,g] = \langle C(f \otimes g_0),g \otimes g_0\rangle$. Then $\|[f,g]\| \leq \|C\| \|f\| \|g\|$, hence there is an operator $w(C)$ in H such that $[f,g] = \langle w(C)f,g\rangle$, $w(I \otimes 1_I)=I$, w is positive linear and $\|w\|= 1$.

$\langle\sum_U A_U U \otimes 1_U(f \otimes g_0), g \otimes g_0\rangle = \sum_U \langle A_U Uf,g\rangle \langle 1_U g_0,g_0\rangle = \langle A_I f,g\rangle$. Thus

$w(\sum_U A_U U \otimes 1_U)=A_I$.

Let r_U denote the right regular representation $(r_U g(V) = g(VU))$ then

$C(f \otimes r_U g_0)=C(I \otimes r_U)f \otimes g_0 = (I \otimes r_U)C(f \otimes g_0)$ (*).

To see that w is faithful note that $w(C^*C)=0$ implies that $C(f \otimes g_o)=0$ and, by (*), $C(f \otimes r_U g_0)=0$.

Since the set of vectors of the form $f \otimes r_U g_0$ generates $H \otimes l^2(\bar{G})$ $C=0$

###

We now claim that $\sum_U A_U U \otimes 1_U=0$ implies $A_U=0$ for all $U \in \bar{G}$. To see this fix V in $\bar{G}$,

$$\sum_U VA_U V^*VU \otimes 1_{VU}=(V \otimes 1_V) \sum_U A_U U \otimes 1_U=0.$$

Hence $VA_{V^*}V^*=w(\sum_U VA_U V^*VU \otimes 1_{VU})=0$ and so $A_{V^*}=0$.

We can now define $\Pi_0:C_0 \to B(\mathcal{a},G)$ by $\Pi_0(\sum_U A_U U \otimes 1_U)= \sum_U A_U U$. Π_0 is a $*$-homomorphism of C_0 onto a dense $*$-subalgebra of B, and by the definition of w,Φ, $\Phi \circ \Pi_0=w$ on C_0.

<u>Lemma 2.6.:</u> Π_0 is bounded if and only if Φ is faithful.

Proof:

1) Suppose Π_0 is bounded, then we can extend it to a map $\Pi: \mathcal{C} \to \mathcal{B}$ such that $\Phi \circ \Pi = w$. $\Pi(\mathcal{C})$ is closed and containing $\Pi_0(\mathcal{C}_0)$, which is dense in $\mathcal{B}$, hence $\Pi(\mathcal{C}) = \mathcal{B}$.

If A is in $\mathcal{B}$, $A \geq 0$, and $\Phi(A)=0$ then $A=\Pi(C)$ for some $C \in \mathcal{C}$, $C \geq 0$, and $w(C) = \Phi(\Pi(C)) = \Phi(A) = 0$. Since w is faithful, this cannot happen and thus Φ is faithful.

2) Suppose Φ is faithful. Let T be in $\mathcal{C}_0$, $||T|| \leq 1$, and $H=T^*T$ then $0 \leq H \leq I$. Consider the C^*-algebra generated by I and $\Pi_0(H)$. We can identify this algebra with the algebra $C(X)$ where X is the pure state space. Let the function f correspond to $\Pi_0(H)$. If $\Pi_0(H) \not\leq I$ then there is $r > 1$ such that $\omega = \{x \in X: f(x) > r\}$ is non empty.

Let g in $C(X)$ be such that: $0 \leq g \leq 1$, g is 1 at some point of ω and g is 0 on $X \setminus \omega$ (if $X \neq \omega$). Since $g(x) \neq 0$ implies $f(x) \geq r$ we have $gf^n \geq r^n g$ for any n. Let A be the operator corresponding to the function g. Then $A \neq 0$, $0 \leq A \leq I$, $A\Pi_0(H) = \Pi_0(H)A$, $A \in \mathcal{B}$, and $A\, \Pi_0(H^n) = A\, \Pi_0(H)^n \geq r^n A$ for any n.

Since Φ is faithful $\Phi(A) \neq 0$, hence there is a state ρ_1 of $\mathcal{B}$ such that $\rho_1(\Phi(A)) \neq 0$. Let $\rho = \rho_1 \circ \Phi$ then ρ is a state, $\rho \circ \Phi = \rho$ and $\rho(A) \neq 0$.

$$\rho(A\, \Pi_0(H^n))^2 \leq \rho(A^2)\, \rho(\Pi_0(H^n)^2) \leq \rho(I)\, \rho(\Pi_0(H^n)^2) = \rho(\Pi_0(H^{2n})) =$$

$$= \rho \circ \Phi(\Pi_0(H^{2n})) = \rho \circ w(H^{2n}).$$

But $H^{2n} \leq I$ and $w(H^{2n}) \leq w(I) = I$ so $\rho \circ w(H^{2n}) \leq \rho(I) = 1$ thus $\rho(A\, \Pi_0(H^n)) \leq 1$.

Since $r^n A \leq A\, \Pi_0(H^n)$ and $r^n \rho(A) \leq \rho(A\, \Pi_0(H^n)) \leq 1$ for any n, we get $\rho(A) = 0$ which is a contradiction. Therefore $\Pi_0(H) \leq 1$ and

$||\Pi_0(T)||^2 = ||\Pi_0(T)^* \; \Pi_0(T)|| = ||\Pi_0(T^*T)|| \leq 1.$

So $||\Pi_0(T)|| \leq 1$ hence Π_0 is bounded. ###

Proposition 2.7.: If $\bar{G}$ is amenable then Φ is faithful.

Proof: For $T = \sum_U A_U U \otimes 1_U$ in $\mathcal{C}_0$ we will show:

$||\sum_U A_U U|| \leq ||T||$, hence $||\pi_0|| \leq 1$.

Fix $\varepsilon > 0$ and let F be a finite subset of $\bar{G}$ such that $A_U=0$ for $U \notin F$. By part (4) of Lemma 2.4 there exists $g \in l^2(\bar{G})$ such that $||g||=1$ and $|\langle 1_U g,g\rangle -1| \leq \varepsilon$ for each U in F. For each pair of unit vectors f and h in H one has: $\langle T(f \otimes g), h \otimes g\rangle = \sum_U \langle A_U Uf \otimes 1_U g, h \otimes g\rangle =$ $\sum_U \langle A_U Uf,h\rangle\langle 1_U g,g\rangle$. Therefore:

$$|\langle T(f \otimes g), h \otimes g\rangle - \langle \sum_U A_U Uf,h\rangle| \leq \sum_U |\langle A_U Uf,h\rangle|\,|\langle 1_U g,g\rangle -1| \leq \varepsilon \sum_U ||A_U U||.$$

Since $|\langle T(f \otimes g),h \otimes g\rangle| \leq ||T||$, we get:

$|\langle \sum_U A_U Uf,h\rangle| \leq ||T|| + \varepsilon \sum_U ||A_U U||$. Since ε is arbitrary, we have: $|\langle \sum_U A_U Uf,h\rangle| \leq ||T||$. Taking the supremum of the left hand side over all pairs of unit vectors completes the proof. ###

The converse of the last proposition is true under the following condition : If $\sum_U a_U U$ is a positive operator in $\mathcal{g}+\mathcal{g}^*$ and a_U are scalars then $\sum_U a_U \geq 0$. We denote this condition by (+).

Note that if each U in $\bar{G}$ comes from a measure preserving transformation on X (i.e. $(Uf)(x)=f(\varphi^{-1}(x))$ where $m \circ \varphi^{-1}=m$), and $m(X) < \infty$, then (+) is satisfied. To see this, let f_0 be the constant function 1 on X and, for $T \in \mathcal{B}$, let $g(T)$ be $\langle Tf_0,f_0\rangle$. Thus g is a positive linear function and $g(\sum_U a_U U)=m(X)(\sum_U a_U)$.

The proof of the following proposition is different from the one given by Arveson.

Let m be the C^*-algebra generated by the operators U in $\bar{G}$.

<u>Proposition 2.8.:</u> Assuming condition (+), if Φ is faithful $\bar{G}$ is amenable.

<u>Proof:</u> Let ρ_1 be a faithful state on $\mathcal{a}$, and let $\rho = \rho_1 \circ \Phi$ then ρ is faithful state on $\mathcal{B}$ and $\rho(U)=0$ for each U in $\bar{G}$ different from I. Consider now the C^*-algebra $\mathcal{L} \subseteq B(l^2(\bar{G}))$ generated by the operators $\{1_U : U \in \bar{G}\}$. Let g_0 be the characteristic function of I, i.e. $g_0(I)=1$, $g_0(U)=0$ for $U \neq I$. Define a linear functional w on $\mathcal{L}$ by $w(T)=\langle Tg_0,g_0\rangle$;

then $w(I)=1$ and $w(1_U)=0$ for $U\neq I$. w is a state on $\mathcal{L}$ and is faithful because $Tg_0=0$ implies $Tr_Ug_0=r_UTg_0=0$ (where $r_Uf(V)=f(VU)$) and the set $\{r_Ug_0 : U \in \bar{G}\}$ is total in $l^2(\bar{G})$.

Since $\bar{G}$ is $\mathcal{A}^-$independent we can define a $*^-$homomorphism φ from a dense $*^-$subalgebra m_0 of m (namely finite linear combinations of elements of $\bar{G}$) onto a dense $*^-$subalgebra of $\mathcal{L}$ by $\varphi(U)=1_U$ and by linearity.

Now, ρ and w are both faithful states and $w \circ \varphi=\rho|_{m_0}$. Therefore, by [7, Proposition 2.5], we can extend φ to a $*^-$isomorphism from m onto $\mathcal{L}$. Condition (+) now implies (3) of Lemma 2.4 and hence $\bar{G}$ is amenable. ###

Lemma 2.9.: Let $\mathcal{D}(g)$ be the norm closed algebra generated by g.

(1) Φ is multiplicative on $\mathcal{D}(g)$.

(2) If Φ is faithful, $\mathcal{D}(g) \cap \mathcal{D}(g)^*=\mathcal{A}$. Thus $\mathcal{D}(g)$ is triangular.

Proof: For (1) it is enough (from Proposition 2.3) to show that Φ is multiplicative on g and, in fact, it suffices to show: $\Phi(AUBV) = \Phi(AU)\,\Phi(BV)$ (*) for A,B in $\mathcal{A}$ and U,V in G. But if either U or V is different from I then both sides of (*) are 0, and if $U=V=I$ then both sides of (*) are AB.

To prove (2) it is enough to prove that any self adjoint operator T in $\mathcal{D}(g)$ is in $\mathcal{A}$. If T is self adjoint then $\Phi(T)$ is also self adjoint and $(T-\Phi(T))^2$ is positive.

But $\Phi((T-\Phi(T))^2)= \Phi(T^2)- \Phi(T)^2$ and this, by (1), equals 0. Since we assumed that Φ is faithful $T = \Phi(T)$. ###

Lemma 2.10.: If Φ is faithful and T is in $\mathcal{B}$ such that $\Phi(TU)=0$ for each $U \in \bar{G}$, then $T=0$.

Proof: Let A be in $\mathcal{A}$ and U in $\bar{G}$, then $\Phi(TAU)=\Phi(TUU^*AU)=\Phi(TU)U^*AU=0$. Therefore $\Phi(TS)=0$ for any $S \in g+g^*$ and by continuity of Φ, $\Phi(TS)=0$ for any S in $\mathcal{B}$. By setting $S=T^*$ we see that $\Phi(TT^*)=0$ hence $T=0$. ###

Unitary operators acting on $\mathcal{A}$.

Lemma 2.11. : Let V in $\mathcal{G}+\mathcal{G}^*$ be unitary operator such that $V^* \mathcal{A} V=\mathcal{A}$, then V has the form $V = \sum_{i=1}^n A_iU_i$ where $A_i \in \mathcal{A}$, $U_i \in \bar{G}$ and:

(1) $A_i=F_iV'$ where F_i are projections in $\mathcal{A}$ and V' is a unitary operator in $\mathcal{A}$.

(2) $\{F_i\}$ and $\{U_i^*F_iU_i\}$ are two orthogonal families of projections in $\mathcal{A}$.

(3) $A_iA_j=A_iA_j^*=0$ for $i\neq j$, and $A_iA_i^*=F_i$.

(4) $\sum_{i=1}^n F_i = \sum_{i=1}^n U_i^*F_iU_i=I$.

Proof: Fix $j \leq n$; let W be VU_j^* $(=\sum_{i=1}^n A_iU_iU_j^*)$ and let F_j be the projection defined in Lemma 1.1 (i.e. F_j is the maximal projection in $\mathcal{A}$ such that $F_jW=WF_j$ and $F_jW \in \mathcal{A}$). For every $Q \leq I-F_j$ there exists a projection $E \leq Q$ in $\mathcal{A}$ such that $EW^*EW=0$. Let $\{E_a\}_{a\in\Lambda}$ be a maximal orthogonal family of subprojections of Q with this property, then $\sum_{a\in\Lambda}E_a=Q$. In fact, if $Q_1=Q-\sum_{a\in\Lambda}E_a$ then there is a non zero projection $E_1 \leq Q_1$ such that $E_1W^*E_1W=0$. Now $\Phi(W)E_a=E_a$ $\Phi(W)E_a=E_a$ $\Phi(WE_a)=E_a$ $\Phi(WE_aW^*W)=E_aWE_aW^*\Phi(W)=0$, so that $\Phi(W)Q=0$ for each subprojeciton Q of $I-F_j$ in $\mathcal{A}$.

Therefore $A_j(I-F_j) = \Phi(W)(I-F_j)=0$ and $A_j=A_jF_j$. As $F_jW \in \mathcal{A}$ and $F_jW = \sum_{i=1}^n F_jA_iU_iU_j^*$, we have $F_jA_j = F_jW$ and $F_jA_i = 0$ for $i\neq j$ (by the $\mathcal{A}$-independence of $\bar{G}$). Since $A_j = F_jA_j = F_jW$; $A_j^*A_j = A_jA_j^* = F_jWW^*F_j = F_j$ and $F_jF_i = F_jA_iA_i^* = 0$. Thus $\{F_j\}_{j=1}^n$ is an orthogonal family of projections in $\mathcal{A}$ and $A_iA_j = A_iA_j^* = 0$ $i\neq j$, $A_iA_i^* = F_i$ Let $F = \sum_{i=1}^n F_i$ then $(I-F)V=0$ hence $I-F = 0$ so $\sum_{i=1}^n F_i = I$. Note that A_i is $F_iVU_i^*$. Let V' be $\sum_{i=1}^n A_i$. Then $A_i = F_iV'$, and $V'V'^* = V'^*V' = \sum_{i,j=1}^n A_i^*A_j =$
$= \sum_{i,j=1}^n (F_iVU_i^*)^*F_jVU_j^* = \sum_{i=1}^n U_iV^*F_iVU_i^* = \sum_{i=1}^n V^*F_iV = V^*(\sum_{i=1}^n F_i)V = I$, from which (1) follows. Now

$$I = V^*V = \sum_{i=1}^n U_i^*A_i^*A_iU_i = \sum_{i=1}^n U_i^*F_iU_i;$$ which establishes (2) and (4).

###

Lemma 2.12.: Let V be a unitary operator in $\mathcal{G}+\mathcal{G}^*$ such that $V^* \mathcal{A} V = \mathcal{A}$, then V acts freely on $\mathcal{A}$ if and only if $\Phi(V) = 0$.

Proof.: Let j in the proof of Lemma 2.11 be such that $U_j = I$ and then $\Phi(V) = F_j V'$ where F_j is the projection associated with V by Lemma 1.1. Since V acts freely if and only if $F_j = 0$ then it acts freely if and only if $\Phi(V) = 0$. ###

We will now generalize this description to other unitary operators acting on $\mathcal{a}$ (not necessarily in $\mathcal{g}+\mathcal{g}^*$).
For this let $V(\mathcal{a})$ be the set of operators in $B(H)$ that are unitary and satisfy $V^* \mathcal{a} V = \mathcal{a}$.
Let W be in $V(\mathcal{a})$ and U in $V(\mathcal{a}) \cap (\mathcal{g}+\mathcal{g}^*)$. Define $F(W,U)$ to be the projection in $\mathcal{a}$ given by Lemma 1.1 applied to the operator WU^*, i.e. $F(W,U)$ is the maximal projection F in $\mathcal{a}$ such that $FWU^* = WU^*F \in \mathcal{a}$. Let $F(W)$ be $\vee\{F(W,U) : U \in \bar{G}\}$.

Lemma 2.13.: For $W \in V(\mathcal{a})$ and $U,V \in \bar{G}$, $U \neq V$, we have:

(1) $F(W,U)F(W,V) = 0$.

(2) $V^*F(W,V)VU^*F(W,U)U = 0$

(3) $F(W) = \bigvee\{F(W,U) : U \in V(\mathcal{a}) \cap (\mathcal{g}+\mathcal{g}^*)\}$.

Proof:

(1) Let $Q = F(W,V)F(W,U)$ then QWU^*, $QWV^* \in \mathcal{a}$ hence $QW \in \mathcal{a}U \cap \mathcal{a}V$. So $QW = 0$ (by the $\mathcal{a}$-independence of $\bar{G}$).

(2) Let $Q = V^*F(W,V)VU^*F(W,U)U$. Then $VQV^* \leq F(W,V)$ and $UQU^* \leq F(W,U)$. Hence WU^*UQU^* and WV^*VQV^* are in $\mathcal{a}$.

(3) Let T be a unitary operator in $V(\mathcal{a}) \cap (\mathcal{g}+\mathcal{g}^*)$ then $T = \sum_{i=1}^n V'F_iU_i$, where $U_i \in \bar{G}$, V' is a unitary operator in $\mathcal{a}$, and $\{F_i\}$ is an orthogonal family of projections in $\mathcal{a}$ with sum I. By definition, $F(W,T) \in \mathcal{a}$ and $WT^*F(W,T)$ $(= \sum_{i=1}^n WU_iF_iV'^*F(W,T)) \in \mathcal{a}$. Hence $WT^*F(W,T)F_i (=WU_i^*F_iV'^*F(W,T)) \in \mathcal{a}$ for each i; and $WU_i^*F_iF(W,T) \in \mathcal{a}$. Thus $F_iF(W,T) \leq F(W,U_i) \leq F(W)$, and $\sum_{i=1}^n F_iF(W,T) = F(W,T) \leq F(W)$, which establishes (3). ###

We will now prove that

$V(\mathcal{a}) \cap (g+g^*) = V(\mathcal{a}) \cap \mathcal{B}(\mathcal{a},G)$. In fact we will prove the following:

Theorem 2.14.: Let W be an operator in $V(\mathcal{a})$ and $T \in g+g^*$ such that $||W-T||<1/3$. Then if $T = \sum_{i=1}^{n} A_i U_i$, where $A_i \in \mathcal{a}$ and $U_i \in \bar{G}$, we have $\vee\{F(W,U_i): i \leq n\} = I$. In particular this holds for any $W \in V(\mathcal{a}) \cap \mathcal{B}(\mathcal{a},G)$.

Proof: Let E be $I- \vee\{F(W,U_i) : i \leq n\}$ and assume $E \neq 0$. $E \leq I-F(W,U_1)$ hence, by Lemma 1.1, there is a projection E_1 in $\mathcal{a}$, $0 \neq E_1 \leq E$, such that $E_1U_1W^*E_1 = 0$.

$E_1 \leq I-F(W,U_2)$ hence, by the same lemma, there is a projection $E_2 \leq E_1$, $E_2 \neq 0$, and $E_2U_2W^*E_2 = 0$. Repeating the use of Lemma 1.1 n times we get a non zero projection Q such that $QU_iW^*Q = 0$ for any $i \leq n$, hence $QTW^*Q = 0$.

We have $2/3 < ||QW||-||QW-QT|| \leq ||QT|| \leq ||T|| \leq ||W|| + ||W-T|| < 4/3$.

and also $||QTW^*Q-QTT^*Q|| \leq ||TW^*-TT^*|| \leq ||T|| \; ||W^*-T^*|| < 4/3\cdot 1/3 = 4/9$.

Since $QTW^*Q = 0$, $||QTT^*Q|| < 4/9$ (*). But $||QT|| > 2/3$ thus $||QTT^*Q|| = ||QT||^2 > (2/3)^2=4/9$ and this contradicts (*). ###

The theorem shows that the distance between two distinct operators in $\bar{G}$ is at least $1/3$ but, in fact, more than this is true as seen in the following lemma.

Lemma 2.15.: Let W,V be in $V(\mathcal{a}) \cap (g+g^*)$ and assume that there is an operator U in $\bar{G}$ such that $F(W,U) \neq 0$ and $F(V,U) = 0$. Then $||W-V||\geq\sqrt{2}$.

Proof: Note that $V^*W \in V(\mathcal{a}) \cap (g+g^*)$ and that $V^*W \notin \mathcal{a}$; for otherwise V and W have the same action on $\mathcal{a}$, as do WU^* and VU^*. But, by hypothesis, VU^* acts freely on $\mathcal{a}$ while WU^* does not. Thus, we can express V^*W in the form described in Lemma 2.11 as $A_0 + \sum_{i=1}^{n} A_iU_i$ where $A_1U_1 \neq 0$ and $U_i \neq U_j$ if $i\neq j$. We have $(V^*W-I)(V^*W-I)^* = \sum_{i,j=1}^{n} A_iU_iU_j^*A_j^* - \sum_{i=1}^{n} A_iU_i - \sum_{i=1}^{n} U_i^*A_i^* + (I-A_0)(I-A_0^*)$.

From Lemma 2.11, $\sum_{i=0}^{n} A_iA_i^* = I$, $A_0 = F_0V' = V'F_0$, and $A_1 = F_1V'$ where F_0 and F_1 are mutually orthogonal projections in $\mathcal{a}$. Since $A_1U_1 \neq 0$; $F_1 \neq 0$

and $A_0+A_0^*$ annihilates some unit vector. Thus

$$2 \leq ||2I-A_0A_0^*||=||\sum_{i=1}^n A_iA_i^*+(I-A_0)(I-A_0^*)||=||\Phi((V^*W-I)(V^*W-I)^*)||$$
$$\leq ||(V^*W-I)(V^*W-I)||=||V^*W-I||^2=||W-V||^2, \qquad \text{since } ||\Phi||=1. \qquad \#\#\#$$

Distinct elements W,V of $\bar{G}$ satisfy the hypothesis of the preceding lemma; for we may use W as U. In this case, $F(W,W) = I$ and $F(V,W) = 0$.

The next theorem describes those operators in $V(\mathcal{a})$ that satisfy $F(W) = I$.

Theorem 2.16.: The conditions, $W \in V(\mathcal{a})$, $\Lambda \subseteq \bar{G}$, and $\vee\{F(W,U) : U \in \Lambda\} = I$ hold if and only if $W = \sum_{U\in\Lambda} VP_UU$ where this sum is understood in the sense of strong-operator convergence if Λ is infinite, V is a unitary operator in $\mathcal{a}$, and $\{P_U\}$, $\{U^*P_UU\}$ are orthogonal families of projections in $\mathcal{a}$ each with sum I. Moreover, if $W \in V(\mathcal{a})$ and $\vee\{F(W,U) : U \in \Lambda\} = I$, the representation of W as $\sum_{U\in\Lambda} VP_UU$, where V is a unitary operator in $\mathcal{a}$, and $\{P_U\}$, $\{U^*P_UU\}$ are orthogonal families of projections in $\mathcal{a}$ each with sum I, is unique, $P_U = F(W,U)$ and $V = W(\sum_{U\in\Lambda} P_UU)^*$.

Proof: a) assume, first, that $W \in V(\mathcal{a})$ and $\vee\{F(W,U) : U \in \Lambda\} = I$. If U_1 and U_2 are distinct elements of $\bar{G}$ and $F = F(W,U_2)$ then $WU_1^*F = FWU_1^* \in \mathcal{a}$ and $WU_2^*F = FWU_2^* \in \mathcal{a}$. Thus $(FWU_2^*)^*(WU_1^*F) = U_2U_1^*F = FU_2U_1^* \in \mathcal{a}$. But $U_2U_1^*$ $(\in\bar{G}\setminus\{I\})$ acts freely on $\mathcal{a}$; so that $F = 0$. It follows that $\{F(W,U) : U \in \Lambda\}$ is an orthogonal family of projections in $\mathcal{a}$. Since H is separable, $F(W,U) = 0$ for all but a conutable number of U in $\bar{G}$. Enumerate the U in Λ for which $F(W,U) \neq 0$ as $U_1,U_2,\ldots,$ let Q_n be $F(W,U_n)$ and P_n be W^*Q_nW. By definition of $F(W,U_n)$ (see Lemma 1.1), $Q_nWU_n^* = WU_n^*Q_n \in \mathcal{a}$; so that $U_nW^*Q_n \in \mathcal{a}$ and, if $A \in \mathcal{a}$, $Q_nA = (Q_nWU_n^*)A(U_nW^*Q_n) = U_nW^*Q_nAWU_n$. Hence $U_n^*Q_nU_n = W^*Q_nAW$. In particular, with I in place of A, $U_n^*Q_nU_n = P_n$ and $U_nP_n = Q_nU_n$.

Since $\{Q_n\}$ is an orthogonal family; $\{U_nP_nU_n^*\}$ and $\{P_n\}$ $(=\{W^*Q_nW\}=\{U_n^*Q_nU_n\})$ are orthogonal families. By assumption $\sum Q_n = I$, so that $\sum P_n = \sum W^*Q_nW = I$. Now U_nP_n is a partial isometry with initial projection $P_n(=(U_nP_n)^*U_nP_n)$

and final projection $Q_n (=U_nP_n(U_nP_n)^*)$ so that $\sum U_nP_n$ is strong-operator convergent to a unitary operator T, and $T^*Q_nAT = P_nU_n^*Q_nAU_nP_n = W^*Q_nAWP_n = W^*Q_nAQ_nW = W^*Q_nAW$ for each n and each A in $\mathcal{a}$. Summing over n, we have $T^*AT = W^*AW$. Thus $WT^*A = AWT^*$ for each A in $\mathcal{a}$; so that WT^* is a unitary operator V in $\mathcal{a}$.

Hence $W = VT = \sum VU_nP = \sum VQ_nU_n$. Letting P_{U_n} be Q_n for $n=1,2,\ldots$ and P_U be 0 when $F(W,U) = 0$, we arrive at the stated representation of W.

b) Suppose, now, that $W = \sum_{U\in\Lambda} VP_UU$, where V is a unitary operator in $\mathcal{a}$ and $\{P_U\}$, $\{U^*P_UU\}$ are orthogonal families of projections in $\mathcal{a}$ with sum I. Then P_UU is a partial isometry with initial projection U^*P_UU $(=Q_U)$ and final projection P_U; so that $\sum_U P_UU$ is a unitary operator T with adjoint $\sum_U U^*P_U$. With A in $\mathcal{a}$, $T^*AP_UT = U^*AP_UU = U^*AUU^*P_UU \in \mathcal{a}Q_U$.
Thus $T^*AT = \sum_{U\in\Lambda} T^*AP_UT = \sum_{U\in\Lambda} U^*AUQ_U \in \mathcal{a}$. But $W = VT$, with V a unitary operator in $\mathcal{a}$; so that W (and T) are in $V(\mathcal{a})$ and $W^*AW = T^*AT$ for each A in $\mathcal{a}$.

Since $P_UW = VP_UU$, we have $P_UWU^* = VP_U \in \mathcal{a}$. In addition $W = \sum_{U\in\Lambda} VUU^*P_UU$; so that $WU^*P_UU = VUU^*P_UU = VP_UU$, and $WU^*P_U = VP_U = P_UWU^*$. Thus $P_U \leq F(W,U)$. Now $\{F(W,U):U \in \Lambda\}$ is an orthogonal family (as noted at the beginning of this proof); and $\sum P_U = I$. Thus $P_U = F(W,U)$ for all U and $\bigvee\{F(W,U) : U \in \Lambda\} = I$. Hence $W \in V(\mathcal{a})$ and the representation of W is unique. ###

Recall that $\mathcal{D}(\mathcal{g})$ is the norm closure of $\mathcal{g}$.

Corollary 2.17.: If $W \in \mathcal{B}(\mathcal{a},G)$ [resp. $\mathcal{D}(\mathcal{g})$] is a unitary operator in $V(\mathcal{a})$, then $W \in \mathcal{g}+\mathcal{g}^*$ [resp. $\mathcal{g}$].

Proof: This result follows from Theorems 2.14 and 2.16, applied to a finite set of $\bar{G}$ (or G). ###

Corollary 2.18.: Let G_1 and G_2 (acting on $\mathcal{a}$) give rise to the triangular algebras $\mathcal{g}_1$, $\mathcal{g}_2$ respectively and to the C*-algebras $\mathcal{B}_1$, $\mathcal{B}_2$ (and $\mathcal{D}(\mathcal{g}_1)$,

$\mathcal{D}(\mathcal{g}_2))$ respectively. Then $\mathcal{B}_1 = \mathcal{B}_2$ (or $\mathcal{D}(\mathcal{g}_1) = \mathcal{D}(\mathcal{g}_2))$ implies $\mathcal{g}_1 + \mathcal{g}_1^* = \mathcal{g}_2 + \mathcal{g}_2^*$ (or $\mathcal{g}_1 = \mathcal{g}_2$)

Proof: Since

$G_1 \subseteq V(\mathcal{a}) \cap \mathcal{B}_1 = V(\mathcal{a}) \cap \mathcal{B}_2 = V(\mathcal{a}) \cap (\mathcal{g}_2 + \mathcal{g}_2^*)$, we have $\mathcal{g}_1 + \mathcal{g}_1^* \subseteq \mathcal{g}_2 + \mathcal{g}_2^*$. Equality follows from symmetry. Similarly, if $\mathcal{D}(\mathcal{g}_1) = \mathcal{D}(\mathcal{g}_2)$ then $G_1 \subseteq V(\mathcal{a}) \cap \mathcal{D}(\mathcal{g}_1) = V(\mathcal{a}) \cap \mathcal{D}(\mathcal{g}_2) = V(\mathcal{a}) \cap \mathcal{g}_2$; and $\mathcal{g}_1 = \mathcal{g}_2$. ###

We will show that the operators in $V(\mathcal{a})$ that satisfy $F(W) = I$ form a group. For this we will need the following lemma.

Lemma 2.19.: For W_1, W_2 in $V(\mathcal{a})$ and U_1, U_2 in $\bar{G}$, $(U_1F(W_2,U_2)U_1^*)F(W_1,U_1) \leq F(W_1W_2,U_1U_2)$.

Proof: Writing A for $W_2U_2^*F(W_2,U_2)$ we have

$W_1W_2U_2^*U_1^*\ (U_1F(W_2,U_2)U_1^*)F(W_1,U_1) = W_1(W_2U_2^*F(W_2,U_2))\ \ U_1^*F(W_1,U_1)$

$= W_1U_1^*U_1AU_1^*F(W_1,U_1)$.

But $W_1U_1^*U_1AU_1^*F(W_1,U_1) = W_1U_1^*F(W_1,U_1)U_1AU_1^* \in \mathcal{a}$ and

$W_1U_1^*F(W_1,U_1)U_1AU_1^* = F(W_1,U_1)W_1U_1^*U_1W_2U_2^*F(W_2,U_2)U_1^* =$

$F(W_1,U_1)W_1U_1^*U_1F(W_2,U_2)W_2U_2^*U_1^* =$

$U_1F(W_2,U_2)U_1^*F(W_1,U_1)W_1U_1^*U_1W_2U_2^*U_1^* =$

$(U_1F(W(W_2,U_2)U_1^*)F(W_1,U_1)W_1W_2U_2^*U_1^*$. This, by the definition of $F(W_1W_2U_1U_2)$, completes the proof. ###

Proposition 2.20.: The set of the operators W in $V(\mathcal{a})$ such that $F(W) = I$ is a group.

Proof:

a) Let W_1, W_2 be in $V(\mathcal{a})$ and $F(W_1) = F(W_2) = I$. By the previous lemma $(U_1F(W_2,U_2)U_1^*)F(W_1,U_1) \leq F(W_1W_2,U_1U_2)$ for $U_1,U_2 \in \bar{G}$ and

$\bigvee\{(U_1F(W_2,U_2)U_1^*)F(W_1,U_1) : U_1,U_2 \in \bar{G}\} = \bigvee\{U_1F(W_2)U_1^*F(W_1,U_1) : U_1 \in \bar{G}\} =$

$\bigvee\{F(W_1,U_1) : U_1 \in \bar{G}\} = F(W_1) = I$. Therefore

$F(W_1W_2,U_1U_2) = I$, and $F(W_1W_2) = I$.

b) We prove that

$F(W^*,U^*) = U^*F(W,U)U = W^*F(W,U)W$ (*) when $W \in V(\mathcal{a})$ and $U \in \bar{G}$.

The last equality of (*) is a consequence of the proof of Theorem 2.16. (In the notation of that proof, $U_n^*A_nU_n = P_n = W^*Q_nW$.) Thus

$W^*UU^*F(W,U)U = W^*F(W,U)WW^*U = U^*F(W,U)UW^*U = U^*(WU^*F(W,U))^*U \in \mathcal{A}$; and

$$U^*F(W,U)U \leq F(W^*,U^*).$$

This same inequality, with W and U replaced by W^* and U^*, yields

$$UF(W^*,U^*)U^* \leq F(W,U) \quad ;$$

and (*) follows. From (*), we have $F(W^*) = W^*F(W)W$.

Thus $F(W^*) = I$ if $F(W) = I$.

Induced algebras

We will prove in Proposition 2.22, that when $\mathcal{G}$ is an irreducible triangular algebra as constructed, from an ordered semigruop G acting freely and ergodically on a maximal abelian algebra $\mathcal{A}$ in $B(H)$, then $P\,\mathcal{G}\,P$, for a projection P in $\mathcal{A}$, is also an irreducible triangular algebra. The algebra $P\,\mathcal{G}\,P$ will said to be the algebra induced by $\mathcal{G}$ on $P(H)$.

Lemma 2.21.: With E a projection in $\mathcal{A}$, let $\mathcal{G}_1$ be the algebra generated by $\mathcal{A}$ and the operators EUE for $U \in G$. Then: $\text{Lat}\,\mathcal{G}_1 = \{P \in \mathcal{A}: EP=0 \text{ or } EP=E\}$.

Proof: If for P in $\mathcal{A}$, $EP=0$ then $EUEP=0=PEUEP$ and if $E=EP$ then $EUEP=EUE=PEUEP$. Hence $P \in \text{Lat}\,\mathcal{G}_1$. If P is in $\text{Lat}\,\mathcal{G}_1$ then P is in $\mathcal{A}$. Moreover, P satisfies $EUEP=PEUEP$; so that $(I-P)EUEPU^*=0$. But if $EP\neq 0$, then by ergodicity, $\vee\{UEPU^*; U \in G\} = I$; and $(I-P)E=0$.
###

Proposition 2.22. : Let $\mathcal{A}$ and G generate the irreducible triangular algebra $\mathcal{G}$. Then $E\,\mathcal{G}\,E(=\{ETE: T \in \mathcal{G}\})$ is an irreducible triangular algebra, for each projection E in $\mathcal{A}$.

Proof: Let E be a projection in $\mathcal{A}$.

$E\,\mathcal{G}\,E \cap (E\,\mathcal{G}\,E)^* = \mathcal{A}E$ hence, on $E(H)$, $E\,\mathcal{G}\,E$ is a triangular algebra.

Let P be a projection in $\mathcal{A}E$. If $P \in \text{Lat}(E\,\mathcal{G}\,E)$ then, considering P as a projection in $\mathcal{A}$, $P \in \text{Lat}\,\mathcal{G}_1$ (in the notation of the preceding lemma).

Therefore EP=0 or E=EP. Since $P \in \mathcal{a}E$, P=0 or P=E. Hence $E\,\mathcal{g}\,E$ is irreducible.

###

An extension of the order on $\bar{G}$

We have ordered $\bar{G}$ by : $U > V$ if $UV^* \in G$. We shall now extend this order to a larger group of operators in $\mathcal{g} + \mathcal{g}^*$.

Let $M(\mathcal{g})$ (or just M) be the group of all operators of the form AU where A in $\mathcal{a}$ is invertible and $U \in V(\mathcal{a}) \cap (\mathcal{g} + \mathcal{g}^*)$.

Let $N(\mathcal{g})$ (or simply N) be $M(\mathcal{g}) \cap \mathcal{g}$. Note that $N(\mathcal{g})$ is a semigroup.

For T,S in $M(\mathcal{g})$, we will say that T is larger than $S (T > S)$ if there is R in $N(\mathcal{g})$ (equivalently, in $\mathcal{g}$) such that T=SR.

The following lemma describes some of the basic properties of this ordering.

Lemma 2.23.: The relation > on $M(\mathcal{g})$ has the following properties:

(1) $T > S$ if and only if S^*T is in $\mathcal{g}$;

(2) it is transitive;

(3) $T > S$ and $S < T$ implies $S \in T\mathcal{a}$;

(4) it is an extension of the order on $\bar{G}$.

Proof:

(1) $T > S$ if and only if $S^{-1}T \in \mathcal{g}$. But $S^{-1}T=(S^*S)^{-1}S^*T$ and, from the definition of $M(\mathcal{g})$, $(S^*S)^{-1}$ is in $\mathcal{a}$. Thus $S^{-1}T \in \mathcal{g}$ if and only if $S^*T \in \mathcal{g}$.

(2) If $T_1=T_2R_1$ and $T_2=T_3R_2$ then $T_1=T_3R_2R_1$. Since $N(\mathcal{g})$ is a semigroup, the order is transitive.

(3) T=SR and S=TL imply I=LR. So $L=R^{-1}=(R^*R)^{-1}R^* \in \mathcal{g}^*$, hence $L \in \mathcal{g} \cap \mathcal{g}^* = \mathcal{a}$, and $S \in T\mathcal{a}$.

(4) Let U,V be in $\bar{G}$. Then $U > V$ with respect to the order on $\bar{G}$ if and only if $V^*U \in G$. On the other hand, U is greater than V with respect to the order on M if and only if $V^*U \in N(\mathcal{g})$. Since $N(\mathcal{g}) \cap \bar{G}=G$ (by the

$\mathcal{a}$- independence of $\bar{G}$), the two orderings coincide on $\bar{G}$.

###

Remark: The reason for extending the order from $\bar{G}$ to $M(g)$ is that the set $M(g)$ is defined in terms of $\mathcal{a}$ and g and is independent of G.

We will say that $T \in N(g)$ is a smallest element in $N(g)$ if : $\Phi(T)=0$ and for any $S \in N(g)$ such that $\Phi(S)=0$, $S > T$.

If T_1, T_2 are both smallest elements in $N(g)$ then $T_1 > T_2 > T_1$ hence $T_2=T_1A$ where A is in $\mathcal{a}$. This shows that the smallest element in $N(g)$ is essentially unique.

Lemma 2.24.: If G has a minimal element V (i.e. $V < U$ for every U in G, $U \neq I$) then BV is smallest in $N(g)$ for each invertible B in $\mathcal{a}$. Conversely, if T is smallest in $N(g)$ then $T=AU$ where $A \in \mathcal{a}$, and U is minimal in G.

Proof:

a) Let V be minimal in G, and let T be in $N(g)$ such that $\Phi(T)=0$. $T= \sum_{i=1}^{n} A_iU_i$ where A_i in $\mathcal{a}$ and $U_i \in G$, $U_i \neq I$. Therefore $B^{-1}V^*T \in g$, because $V^*U_i \in G$ for all i. Hence $T > BV$.

b) Let $T = \sum_{i=1}^{n} A_iU_i$ be smallest in $N(g)$ and assume that U_1 is the smallest among the U_i's. Since $U_1^*T \in g$; $T > U_1$. But T is smallest. Thus $T=AU_1$, $A \in \mathcal{a}$. If $U \in G\setminus\{I\}$ and $U_1 > U$ then $T > U$ (in $N(g)$) so that $U=CT=CAU_1$ for some C in $\mathcal{a}$. By $\mathcal{a}$-independence of G, $U=U_1$; and U_1 is minimal in G.

###

It is also natural to define $T \in N(g)$ to be minimal if $S < T$ for S in $N(g)$, implies that $S \in \mathcal{a}$ or $S \in \mathcal{a}T$.

Lemma 2.25.:

(1) If T is smallest in $N(g)$ and T acts ergodically on $\mathcal{a}$ then T is minimal.

(2) If T is smallest in $N(g)$ and S in $N(g)$ satisfies $S < T$, then

$S=A+BT$ where $A,B \in \mathfrak{a}$.

(3) If T is minimal in $N(\mathfrak{g})$ and $\Phi(T)=0$, then T is smallest.

Proof:

(2) Let T be smallest in $N(\mathfrak{g})$ and S in $N(\mathfrak{g})$ satisfy $S < T$. From the previous lemma $T=AU$, where U is minimal in G. Let $S=A_0+A_1U_1+\ldots+A_nU_n$ then $S^*T=(A_0^*+U_1^*A_1^*+\ldots+U_n^*A_n^*)AU \in \mathfrak{g}$. By the $\mathfrak{a}$-independence of $\bar{G}$, $U_i^*U \in G$ for $i=1,2,\ldots,n$.

Hence, for each $i \leq n$, $U > U_i$. Since U is minimal in G, $U_i=I$ or $U_i=U$ and $S=A_0+A_1U$.

(1) Let T be a smallest element in $N(\mathfrak{g})$ that acts ergodically on $\mathfrak{a}$. We know that $T=AU$ where U is minimal in G, and, by (2), to prove minimality we should show that $D+BU \in N(\mathfrak{g})$ $(D,B \in \mathfrak{a})$ implies $D=0$ or $B=0$.

Suppose $S=D+BU \in N(\mathfrak{g})$. Then, by Lemma 2.11, $S=C(F+EU)$ where $C \in \mathfrak{a}$ and F,E are projections in $\mathfrak{a}$ satisfying: $FE=0$, $F+E=I$, and $F+U^*EU=I$. Hence $E=I-F=U^*EU$. Ergodicity of T (and thus of U) implies that $E=0$ or $E=I$.

(3) Let T be minimal in $N(\mathfrak{g})$ satisfying $\Phi(T)=0$. We can write $T=\sum_{i=1}^{n} A_iU_i$. Let U_1 be the smallest among the U_i's. Since $U_1^*U_i \in G$ for each i, $U_1^*T \in \mathfrak{g}$. Thus $T > U_1$ and, by minimality of T, $T \in \mathfrak{a}U_1$ and U_1 is minimal in G. By the previous lemma T is smallest in $N(\mathfrak{g})$.

###

The following definitions are based on the terminology of [8,Chapter XIII].

Definitions:

(1) G is integrally closed if, for U and V in G, $U^n < V$ for $n=1,2,\ldots$ implies $U=I$.

(2) G is complete if any bounded subset of $\bar{G}$ has a l.u.b. and a g.l.b.

(3) G satisfies the (DCC) condition if every non empty subset of G contains a minimal member.

Lemma 2.26.: If G has a minimal element U, then the following are equivalent:

(1) $G = \{U^n : n=0,1,2,\ldots\}$.

(2) G is integrally closed.

(3) G is complete.

(4) G satisfies the (DCC) condition.

Proof: We will prove (3) implies (2), (2) implies (1), and (4) implies (1). As (1) clearly implies (3) and (4), this will complete the proof. (3) implies (2): This part of the lemma is true also if G does not have a minimal element, and is proved in [8, Chapter XIII, p.291, Lemma 5]. Let W,V in G be such that $W^n < V$ for all $n \geq 1$.

By completness, the set $\{W^n : n=1,2,\ldots\}$ has a l.u.b. W_0. W_0W is therefore the l.u.b. of the set $\{W^{n+1} : n=1,2,\ldots\}$ and, since the latter set is contained in the former, $W_0W < W_0$. But $W_0W > W_0$ implies $W=I$.

(2) implies (1) : Let V be an element of G different from I. Since G is integrally closed, there exists n such that $V < U^n$. Let m be the minimal positive integer satisfying this, then $U^{m-1} < V < U^m$ and $U^{m-1} \neq V$. Hence $U > VU^{-m+1} > I$, $VU^{-m+1} \neq I$, and by minimality of U this implies $V=U^m$.

(4) implies (1) : Let V be an element of G different from I, and consider the set of all operators in G of the form VU^{-n}, where n is a positive integer. Since $U < V$, this set contains VU^{-1} and is nonempty.

Since G satisfies the (DCC) condition, this set has a minimal member, say VU^{-m}. We get $U^m < V < U^{m+1}$, $U^{m+1} \neq V$, and, as before, this implies $V=U^m$.

###

Proposition 2.27.: G is integrally closed if and only if for any $S,T \in N(\mathcal{g})$ such that $\Phi(S)=0$ there exists $n \geq 1$ such that $S^n > T$.

Proof.: With U and V in place of S and T the given condition implies at once that G is integrally closed.

Suppose, now, that $S,T \in N(\mathcal{g})$ and that $\Phi(S)=0$. We can write $S = A_1U_1+\ldots+A_nU_n$ and $T = B_0+B_1V_1+\ldots+B_mV_m$ where $A_i,B_j \in \mathcal{a}$ and $U_i,V_j \in G\setminus\{I\}$. Assume that U_1 is the smallest among the U_i's and V_m is the greatest among the V_j's. By the assumption that G is integrally closed, there is a positive integer k such that $U_1^k > V_m$. S^k is a sum of terms each of the form CW where $C \in \mathcal{a}$ and W, in G, is greater than U_1^k, thus $S^k > U_1^k$. By choice of V_m, $T^*V_m \in \mathcal{g}$ and from Lemma 2.23 (1) $V_m > T$. Since $U_1^k > V_m > T$, we get $S^k > T$. ###

If $G = \{U^n : n = 0,1,2,\ldots\}$, for some operator U, we say that G is singly generated.

Corollary 2.28. : $\mathcal{g}$ is generated by a singly generated semigroup if and only if $\mathcal{g}$ satisfies the following two conditions:

(1) $N(\mathcal{g})$ has a smallest element.

(2) For any S,T in $N(\mathcal{g})$ such that $\Phi(S) = 0$ there exists $n \geq 1$ such that $S^n > T$.

Proof: By Lemma 2.24, (1) is equivalent to the existence of a minimal element in G. By proposition 2.27, (2) is equivalent to the property: G is integrally closed. Lemma 2.26 completes the proof. ###

Remark: Conditions (1) and (2) of the previous corollary are independent of the choice of G. They depend only on $\mathcal{g}$ and $\mathcal{a}$.
To see this it should be recalled that for S in $M(\mathcal{g})$, $\Phi(S) = 0$ if and only if S acts freely on $\mathcal{a}$ (Lemma 2.12).

Theorem 2.29.: Let G_1 and G_2 be two ordered semigroups acting freely and ergodically on the maximal abelian algebra $\mathcal{a}$ and generating the same irreducible triangular algebra $\mathcal{g}$.
If G_1 is singly generated, with generator U, then G_2 is also singly generated and its generator is in $\mathcal{a}U$.

Proof: The fact that G_2 is also singly generated is a result of Corollary 2.28. Let V be the generator of G_2, then V and U are both smallest elements in $N(\mathcal{g})$, hence, by Lemma 2.24, $V \in \mathcal{a}U$. ###

Proposition 2.30: If $\mathcal{g}$ is generated by the semigroup G and satisfies conditions (2) of Corollary 2.28, then G is abelian.

Proof: By Proposition 2.27 G is integrally closed. By [8, Theorem 12, Chapter XIII] this implies that G is isomorphic to a subsemigroup of the non negative real numbers (under addition), hence abelian. ###

CHAPTER III: THE HOMOMORPHISMS OF $\mathcal{G}$

In this chapter we deal, once again, with the triangular algebra $\mathcal{G}$ generated by a maximal abelian algebra $\mathcal{A}$, acting on a separable Hilbert space H, and an ordered semigroup of unitary operators. We will study the linear multiplicative maps from $\mathcal{G}$ (or $\mathcal{G} + \mathcal{G}^*$) into the algebra B(K) of all bounded operators on some Hilbert space K, whose restriction to $\mathcal{A}$ is self adjoint. In particular we will deal with such automorphisms of $\mathcal{G}$.

Definitions:

(1) $\Psi: \mathcal{G} \longrightarrow B(K)$ is a homomorphism if it is multiplicative, linear, and $\Psi(A^*)=\Psi(A^*)$ for $A \in \mathcal{A}$.

(2) A homomorphism Ψ is a bounded homomorphism if

$$\text{Sup}\ \{||\Psi(T)|| : ||T|| \leq 1\} < \infty.$$

(3) A $*$-homomorphism is a homomorphism Ψ which satisfies $\Psi(U)^* = \Psi(U)^{-1}$ for each U in G.

(4) Isomorphisms and $*$-isomorphisms are one-to-one homomorphisms and $*$-homomorphisms respectively.

Proposition 3.1.: Let $\Psi: \mathcal{G} \longrightarrow B(K)$ be a homomorphism, then Ψ can be extended to a $*$-homomorphism of $\mathcal{G}+\mathcal{G}^*$ if and only if Ψ is a $*$-homomorphism If Ψ is a bounded $*$-homomorphism then it can be extended to $\mathcal{B}(\mathcal{A},G)$.

Proof: It is clear that, if Ψ can be extended to a $*$-homomorphism Ψ' on $\mathcal{G}+\mathcal{G}^*$ then, Ψ is a $*$-homomorphism on $\mathcal{G}$.
For the converse, suppose that $\Psi(U)$ is a unitary operator for each U in G. Each operator T in $\mathcal{G}+\mathcal{G}^*$ can be expressed in only one way (by $\mathcal{A}$- independence of $\bar{G}$) in the form $A_0+A_1U_1+\ldots+A_nU_n$, where $A_j \in \mathcal{A}$ and $U_j \in \bar{G}$. Define $\Psi'(T)$ to be $\Psi(A_0)+\Psi(A_1)\Psi'(U_1)+\ldots+\Psi(A_n)\Psi'(U_n)$, where $\Psi'(U_j) = \Psi(U_j)$ if $U_j \in G$ and $\Psi'(U_j) = \Psi(U_j^*)^{-1}$ if $U_j^* \in G$. We have that

Ψ' is well defined and linear on $\mathcal{g}+\mathcal{g}^*$.

To see that Ψ' is multiplicative, it will suffice to show that $\Psi'(AUBV) = \Psi(A)\Psi'(U)\Psi(B)\Psi'(V)$ when $A,B \in \mathcal{a}$ and $U,V \in \bar{G}$. Suppose, for example, that $U \notin G$, $V \in G$, and $UV \in G$. Then $U^* \in G$ and

$\Psi'(V) = \Psi(V) = \Psi(U^*UV) = \Psi(U^*)\Psi(UV)$; so that

$\Psi'(UV) = \Psi(UV) = \Psi(U^*)^{-1}\Psi(V) = \Psi'(U)\Psi'(V)$.

We also have, $\Psi(BU^*) = \Psi(U^*UBU^*) = \Psi(U^*)\Psi(UBU^*)$; so that

$\Psi'(U)\Psi'(B)\Psi'(U^*) = \Psi(U^*)^{-1}\Psi(B)\Psi(U^*) = \Psi(U^*)^{-1}\Psi(BU^*) = \Psi(UBU^*)$. Thus

$$\Psi'(AUBV) = \Psi'(AUBU^*UV) = \Psi(A)\Psi(UBU^*)\Psi'(UV) = \Psi(A)\Psi'(U)\Psi(B)\Psi'(U^*)\Psi'(U)\Psi'(V) = \Psi(A)\Psi'(U)\Psi(B)\Psi'(V).$$

The other cases for U,V, and UV, are handled by similar computations. Thus Ψ' is a *-homomorphism of $\mathcal{g}+\mathcal{g}^*$ that extends Ψ.

For the last assertion of this proposition, note that, with T in $\mathcal{g}+\mathcal{g}^*$ $TU \in \mathcal{g}$ for a suitable U in $\bar{G}$. If $||T||(=||TU||) \leq 1$ then $||\Psi'(T)|| = ||\Psi'(T)\Psi'(U)|| = ||\Psi'(TU)|| = ||\Psi(TU)|| \leq ||\Psi||$. Thus $||\Psi'|| = ||\Psi||$ and, in particular, Ψ' is bounded when Ψ is . ###

We will mostly deal with homomorphisms satisfying $\Psi(\mathcal{a}) = \Psi(\mathcal{a})'$ (i.e.the image of Ψ is maximal abelian).

<u>Proposition 3.2.:</u> Each homomorphism $\Psi : \mathcal{g} \longrightarrow B(K)$ whose restriction to $\mathcal{a}$ is weak operator continuous and for which $\Psi(\mathcal{a}) = \Psi(\mathcal{a})'$ has the form $\Psi = \Psi_2 \cdot \Psi_1$ where:

a) Ψ_1 is a homomorphism from $\mathcal{g}$ into the algebra $F\mathcal{g}F$ (with F some projeciton in $\mathcal{a}$) satisfying: $\Psi_1(A)=AF$ for A in $\mathcal{a}$ and $\Psi_1(U)=F\varphi(U)UF$ for U in G, where φ maps G into $\mathcal{a}F$.

b) $\Psi_2 : F\mathcal{g}F \longrightarrow B(K)$ is implemented by a unitary transformation $W:K \longrightarrow F(H)$ i.e. $\Psi_2(T)=W^*TW$.

<u>Proof:</u> Since $\Psi|\mathcal{a}$ is weak operator continuous, its kernel has the form $\mathcal{a}(I-F)$ for some projection F in $\mathcal{a}$, and $\Psi|\mathcal{a}F$ is a *-isomorphism onto $\Psi(\mathcal{a})$. Since $\mathcal{a}F$ is maximal abelian on $F(H)$ and $\Psi(\mathcal{a})$ is maximal abelian

(by assumption); $\Psi|\mathcal{a}F$ is implemented by a unitary transformation $W:K \longrightarrow F(H)$ (i.e. $\Psi(AF) = W^*AFW$ for A in $\mathcal{a}$).

Consider $\Psi_1:\mathcal{g} \longrightarrow B(F(H))$ defined by $\Psi_1(S)=W\Psi(S)W^*$ (so that $\Psi_1(A)=AF$ for A in $\mathcal{a}$). If $A \in \mathcal{a}$ and $U \in G$, then $\Psi_1(A)\,\Psi_1(U) = \Psi_1(U)\Psi_1(U^*AU)$ so that $AF\Psi_1(U) = \Psi_1(U)U^*AUF$. But $\Psi_1(U) = F\Psi_1(U)F$.

Hence $AF\Psi_1(U)U^* = F\Psi_1(U)U^*A$ so that $F\Psi_1(U)U^* \in \mathcal{a}' = \mathcal{a}$.

Let $\varphi(U)$ be $F\Psi_1(U)U^*$. Then $F\Psi_1(U) = F\varphi(U)U$ and $F\Psi_1(U) = F\varphi(U)UF$.

Hence $\Psi_1(U) = F\varphi(U)UF \in F\mathcal{g}F$. ###

Proposition 3.3.: Let F be a projection in $\mathcal{a}$, $F\neq I$, and let $\Psi:\mathcal{g} \longrightarrow F\mathcal{g}F$ be a homomorphism such that $\Psi(A) = AF$ for A in $\mathcal{a}$ and $\Psi(U) = F\varphi(U)UF$ for U in G where $\varphi(U)$ is in $\mathcal{a}F$.

Then there exists some U in G such that $\varphi(U)$ has a non zero null space in $F(H)$. Thus if Ψ satisfies the conditions of Proposition 3.2 and its restriction to $\mathcal{a}$ is not an isomorphism then there is some U in G such that $\Psi(U)$ has a non zero null space.

Proof: As G acts ergodically on $\mathcal{a}$, there is a V in G such that $Q=FV^*(I-F)V\neq 0$. Since $VQV^* \leq I-F$, $FVQV^*=0$ and $FVA=0$. As G acts ergodically on $\mathcal{a}$, and $VQV^*\neq 0$ there is some W in G such that $FWVQV^*W^*\neq 0$ hence $FWVQ\neq 0$. If $\varphi(WV)$ annihilates no vector in $F(H)$ then $0 \neq \varphi(WV)FWVQ = \varphi(WV)FWVFQ = \varphi(WV)Q = \varphi(W)\,\varphi(V)Q = \Psi(W)\,\varphi(V)FVFQ = 0$ (since $FVQ=0$) - a contradiction. Therefore $\varphi(WV)$ annihilates some non zero vector in $F(H)$. With WV for U, the first assertion follows.

The last assertion follows from the fact that $F\neq I$ if $\Psi|\mathcal{a}$ is not an isomorphism and $\Psi(U) = \varphi(U)UF = \varphi(U)FUF$. ###

Our interest is, primarily, in the case where F (above) is I. The next definition refers to this case.

Definition: Let $W:K \longrightarrow H$ be a unitary transformation and let φ be a map from G(resp. $\bar{G}$) into $\mathcal{a}$.

A homomorphism Ψ from $\mathcal{g}$ (resp. $\mathcal{g}+\mathcal{g}^*$) into $B(K)$ will be said to be implemented by (W,φ) if the following hold:

For $A \in \mathfrak{a}$, $\Psi(A) = W^*AW$.

For $U \in \bar{G}$, $(U) = W^*\varphi(U)UW$.

If Ψ is a homomorphism on $\mathfrak{g}+\mathfrak{g}^*$ (resp. $\mathfrak{g}$) implemented by (W,φ) then for all U,V in $\bar{G}$ (resp. G) $\varphi(UV) = W\Psi(UV)W^*V^*U^* = W\Psi(U)\Psi(V)W^*V^*U^* =$

$W\Psi(U)W^*U^*UW\Psi(V)W^*V^*U^*$. Hence

$$\varphi(UV) = \varphi(U)U\varphi(V)U^* \qquad (*).$$

Conversly, if (*) holds, for all U,V in $\bar{G}$ (resp. G), and $W:K \longrightarrow H$ is a unitary transformation, then there is a homomorphism Ψ from $\mathfrak{g}+\mathfrak{g}^*$ (resp. $\mathfrak{g}$) into $B(K)$ which is implemented by (W,φ).

In fact, each operator T in $\mathfrak{g}+\mathfrak{g}^*$ (resp. $\mathfrak{g}$) can be expressed in only one way (by $\mathfrak{a}$-independence of $\bar{G}$) in the form $A_0+A_1U_1+\ldots+A_nU_n$, where $A_i \in \mathfrak{a}$ and $U_i \in \bar{G}$ (resp. G). Define $\Psi(T)$ to be $W^*(A_0+A_1\varphi(U_1)U_1+\ldots+A_n\varphi(U_n)U_n)W$.

We have that Ψ is well defined and linear on $\mathfrak{g}+\mathfrak{g}^*$ (resp. $\mathfrak{g}$).

To see that Ψ is multiplicative, it will suffice to show that $\Psi(AUBV) = \Psi(AU)\Psi(BV)$ when $A,B \in \mathfrak{a}$ and $U,V \in \bar{G}$ (resp. G) But

$\Psi(AUBV) = \Psi(AUBU^*UV) = W^*AUBU^*\ (UV)UVW = W^*AUBU^*\ (U)U\ (V)U^*UVW =$

$W^*A\varphi(U)UBU^*U\varphi(V)VW = W^*A\varphi(U)UWW^*B\varphi(V)VW$

Thus $\Psi(AUBV) = \Psi(AU)\Psi(BV)$.

Hence Ψ is a homomorphism and is implemented by (W,φ).

We will now turn to the case where Ψ is an isomorphism.

Lemma 3.4: Let Ψ be a homomorphism of $\mathfrak{g}$ satisfying $\Psi(\mathfrak{a}) = \Psi(\mathfrak{a})'$ and such that $\Psi|\mathfrak{a}$ is an isomorphism, then Ψ is implemented by (W,φ) for some W and φ.

Proof: Apply Proposition 3.2 in the case where $F=I$. ###

Proposition 3.5: Let Ψ be a homomorphism of $\mathfrak{g}$ into $B(K)$ satisfying $\Psi(\mathfrak{a}) = \Psi(\mathfrak{a})'$ then the following are equivalent:

(1) Ψ is an isomorphism.

(2) Ψ is implemented by some (W,φ) where $\varphi(U)$ is invertible for each U in G.

(3) $\Psi|\mathcal{a}$ is weak operator continuous and $\Psi(U)$ is invertible for any U in G.

(4) $\Psi|\mathcal{a}$ is a *-isomorphism and $\Psi(U)$ is invertible for each U in G.

Proof:

(1) implies (2): Assume that Ψ is an isomorphism. By Lemma 3.4, Ψ is implemented by some (W,φ). If, for some U in G, $\varphi(U)$ is not invertible, then there exists a non zero projection F in $\mathcal{a}$ such that $\varphi(U)F=0$. In this case: $\Psi(FU) = W^*FWW^*\varphi(U)UW = W^*F\varphi(U)UW = 0$ contradicting the assumption that Ψ is an isomorphism.

(2) implies (1): Assume that (2) holds and that for some $A_i \in \mathcal{a}$ and $U_i \in G$, $\Psi(A_0+A_1U_1+\ldots+A_nU_n) = 0$. Then $0 = W^*(A_0+A_1\varphi(U_1)U_1+\ldots+A_n\varphi(U_n)U_n)W$. Thus $0 = A_0+A_1\varphi(U_1)U_1+\ldots+A_n\varphi(U_n)U_n$ and so $A_i = 0$ for each i.

(2) implies (3): Since $\Psi|\mathcal{a}$ is implemented by a unitary operator, and $\Psi(U) = W^*\varphi(U)UW$; $\Psi|\mathcal{a}$ is weak operator continuous and $\Psi(U)$ is invertible.

(3) implies (4): Assuming $\Psi|\mathcal{a}$ is weak operator continuous, we can apply Proposition 3.2. If $\Psi|\mathcal{a}$ is not an isomorphism then $F\neq I$, hence, by Proposition 3.3, there is a U in G such that $\Psi(U)$ is not invertible contradicting the assumption of (3).

(4) implies (2) : This is the assertion of Lemma 3.4. ###

In Proposition 3.5 we assumed that the image of $\Psi|\mathcal{a}$ is a maximal abelian algebra. If we assume only that the commutant of the image is of type $I_m (m \leq \infty)$ we get the following:

Proposition 3.6.: Let $\Psi: \mathcal{g} \longrightarrow B(K)$ be an isomorphism such that $\Psi(\mathcal{a})'$ is of type $I_m (m \leq \infty)$, then Ψ is unitarily equivalent to a homomorphism τ of the following form:

$\tau: \mathcal{g} \longrightarrow B(H\otimes K_m)$ (where $\dim K_m=m$) and $\tau(A)=A\otimes I_m$ for A in $\mathcal{a}$,

$\tau(U) = C(U)(U\otimes I_m)$ for U in G, where $C(U)$ is in $\mathcal{a}\otimes B(K_m)$ and I_m is the identity operator on K_m.

Proof: Define $w:\mathcal{g} \longrightarrow \mathcal{g}\otimes I_m$ by $w(S) = S\otimes I_m$ for S in $\mathcal{g}$. Then $w(\mathcal{a})' = (\mathcal{a}\otimes I_m)' = \mathcal{a}\otimes B(K_m)$ is a type I_m von Neumann algebra. The map $\Psi\cdot w^{-1}$ is a *-isomorphism from $w(\mathcal{a})$ onto $\Psi(\mathcal{a})$, two algebras of type I_1 with commutants of type I_m. Hence $\Psi\cdot w^{-1}$, restricted to $\mathcal{a}\otimes I_m$, is implemented by a unitary transformation $V:K \longrightarrow H\otimes K_m$.

With U in G and A in $\mathcal{a}$, $\Psi(A)\Psi(U) = \Psi(U)\Psi(U^*AU)$; so that $V^*(A\otimes I_m)V\Psi(U) = \Psi(U)V^*(U^*AU\otimes I_m)V$.

Hence $(A\otimes I_m)V\Psi(U)V^*(U^*\otimes I_m) = V\Psi(U)V^*(U^*\otimes I_m)(A\otimes I_m)$. Let $C(U)$ be $V\Psi(U)V^*(U^*\otimes I_m)$ then $C(U) \in (\mathcal{a}\otimes I_m)' = \mathcal{a}\otimes B(K_m)$ and $\Psi(U) = V^*C(U)(U\otimes I_m)V$. By choice of V, $\Psi(A) = V^*(A\otimes I_m)V$ for A in $\mathcal{a}$. Thus defining $\tau(S)$ to be $V\Psi(S)V^*$, completes the proof. ###

Corollary 3.7: If $\Psi:\mathcal{g} \longrightarrow B(K)$ is a homomorphism such that $\Psi(\mathcal{g})\cap\Psi(\mathcal{g})^*$ is abelian and $\Psi(\mathcal{a}) = \Psi(\mathcal{a})'$, then conditions (1)-(4) of Proposition 3.5 are equivalent, and if any one of them is fulfilled, $\Psi(\mathcal{g})$ is an irreducible triangular algebra unitarily equivalent to $\mathcal{g}$.

Proof: Since $\Psi(\mathcal{a})$ is maximal abelian and contained in the abelian algebra $\Psi(\mathcal{g})\cap\Psi(\mathcal{g})^*$; $\Psi(\mathcal{g})\cap\Psi(\mathcal{g})^* = \Psi(\mathcal{a})$, and $\Psi(\mathcal{g})$ is triangular. From Proposition 3.5, Ψ is implemented by some (W,φ). As $\varphi(U)$ is invertible for each U in G and $W^*AW = \Psi(A)$ for each A in $\mathcal{a}$; $W^*UW = \Psi(\varphi(U)^{-1}U)$. Thus the mapping, $T \longrightarrow W^*TW$, of into $B(K)$ maps $\mathcal{g}$ in $\Psi(\mathcal{g})$. Now $\Psi(A)\Psi(U) = W^*A\varphi(U)UW$ when $A \in \mathcal{a}$ and $U \in G$; so that the mapping, $T \longrightarrow W^*TW$, carries $\mathcal{g}$ onto $\Psi(\mathcal{g})$. It follows that $\Psi(\mathcal{g})$ is an irreducible triangular algebra unitary equivalent to $\mathcal{g}$. ###

Corollary 3.8.: (1) Any *-homomorphism $\Psi:\mathcal{g} \longrightarrow B(K)$ that maps $\mathcal{a}$ onto a maximal abelian algebra is a *-isomorphism if and only if $\Psi|\mathcal{a}$ is an isomorphism.

(2) Any automorphism of $\mathcal{g}$ (i.e. isomorphism from $\mathcal{g}$ onto itself) that maps

$\mathcal{A}$ onto iteself is implemented by some (W,φ) where φ is a map form G into $\mathcal{A}$, $\varphi(U)$ is invertible for each U in G, and W is in $V(\mathcal{A})$.

Proof: (1) By the definition of a *-homomorphism, $\Psi(U)$ is a unitary operator for each U in G. Thus Ψ is a *-isomorphism by the equivalence of conditions (1) and (4) of Proposition 3.5.

For (2) apply the equivalence of conditions (1) and (2) of Proposition 3.5 for the case where $\Psi(\mathcal{A}) = \mathcal{A}$. As $W^*\mathcal{A}W = \Psi(\mathcal{A}) = \mathcal{A}$, $W \in V(\mathcal{A})$.

###

Proposition 3.9.: Let $\mathcal{g}_i$ $(i=1,2)$ be irreducible triangular algebras generated by $\mathcal{A}_i$ and G_i and $\mathcal{B}_i$ the C*-algebras they generate. Let $\Psi:\mathcal{B}_1 \longrightarrow \mathcal{B}_2$ be a *-homomorphism satisfying $\Psi(\mathcal{A}_1) = \mathcal{A}_2$. Then $\Psi(\mathcal{g}_1+\mathcal{g}_1^*) \subseteq \mathcal{g}_2 + \mathcal{g}_2^*$.

Proof: Let U be a unitary operator in $\bar{G}_1$. Then $\Psi(U)$ is unitary in $\mathcal{B}_2$ and $\Psi(U)^*\Psi(A)\Psi(U) = \Psi(U^*AU) \in \Psi(\mathcal{A}_1)$ for A in $\mathcal{A}_1$.
Hence $\Psi(U)$ is in $V(\mathcal{A}_2)\cap\mathcal{B}_2$, and by Corollary 2.17, $\Psi(U)$ is in $\mathcal{g}_2+\mathcal{g}_2^*$. So $\Psi(\bar{G}_1) \subseteq \mathcal{g}_2+\mathcal{g}_2^*$, and by assumption $\Psi(\mathcal{A}_1) \subseteq \mathcal{g}_2+\mathcal{g}_2^*$. Hence $\Psi_1(\mathcal{g}_1+\mathcal{g}_1^*) \subseteq \mathcal{g}_2+\mathcal{g}_2^*$.

###

States and representations. The GNS construction.

Definitions:

(1) $f : \mathcal{g} \longrightarrow \mathbb{C}$ is a state on $\mathcal{g}$ if f is linear, $f|\mathcal{A}$ is a state on $\mathcal{A}$, and $||f|| \leq 1$.

(2) By a representation of $\mathcal{g}$ on the Hilbert space K we mean a bounded homomorphism from $\mathcal{g}$ into $B(K)$.

We will construct a representation from a given state, using the GNS construction, and establish the uniqueness of a cyclic representation that gives rise to a given state.

Remark: Any state of $\mathcal{g}$ has a unique extension to a state of $\mathcal{B}(\mathcal{A},G)$. In fact, if f is a state of $\mathcal{g}$ it has an extension g of norm 1 to $B(H)$ (from the Hahn-Banach Theorem). Since $I \in \mathcal{A}$ and $f|\mathcal{A}$ is a state of

$\mathcal{a}$, f(I)=1=g(I). Thus g is a state of B(H) and the restriction of g to $\mathcal{B}(\mathcal{a},G)$ is the unique extension of f to a state of $\mathcal{B}(\mathcal{a}, G)$

Proposition 3.10.: To any state g of $\mathcal{B}(\mathcal{a},G)$ we can associate two representations, a representation Π of $\mathcal{B}(\mathcal{a},G)$ on a Hilbert space K, and a representation Π' of g on a Hilbert space K', and vectors $\xi \in K$, $\xi' \in K'$ such that : (1) $\langle \Pi(T)\xi,\xi\rangle = g(T)$ $\quad T \in \mathcal{B}(\mathcal{a},G)$.

$$\langle \Pi'(T)\xi', \xi'\rangle = g(T), \qquad T \in g.$$

(2) $K=[\Pi(\mathcal{B}(\mathcal{a},G))\xi]$, $K'=[\Pi'(g)\xi']$ where [M] denotes the closed linear subspace generated by M. (3) There is a unique isometric linear mapping τ of K' into K such that, for each T in g, $\Pi(T)\cdot\tau = \tau\cdot\Pi'(T)$ and $\tau(\xi')=\xi$.
(4) For each U in G, $\Pi'(U)$ is an isometry.

Proof: Let Π, ξ, and K be the representation, the cyclic vector and the Hilbert space obtained by the GNS construction applied to the state g of $\mathcal{B}(\mathcal{a},G)$.

Let N be $\{T \in g : g(T^*T)=0\}$. Then N is a left ideal in g. Let K_0 be the linear space g/N equipped with the inner product $\langle S+N,T+N\rangle = g(T^*S)$; and let K' denote the completion of K_0. Then K' is a Hilbert space.

Define a map $\tau': K_0 \longrightarrow K$ by $\tau'(T+N)=T+N_1$ where $N_1 = \{T \in \mathcal{B}(\mathcal{a},G) : g(T^*T) = 0\}$. τ' is certainly an isometry and can be extended to an isometry map, τ, from K' into K. Since $\langle TS+N,TS+N\rangle = g(S^*T^*TS) \leq ||T||^2 \langle S+N,S+N\rangle$; (*) we can define an operator $\Pi'(T)$ on K_0 by $\Pi'(T)(S+N)=TS+N$ and extend it to a bounded operator on K'.

Clearly, $T \longrightarrow \Pi'(T)$ is a representation of g (by (*) $||\Pi'(T)|| \leq ||T||$).

Let ξ' be the vector I+N in K_0 and, since ξ is $I+N_1$, $\tau(\xi')=\xi$. Now $g(T)=g(IT)=\langle T+N,I+N\rangle = \langle \Pi'(T)\xi', \xi'\rangle$ for $T \in g$ so this proves (1).

For (2), note that $K'=[\Pi'(g)\xi']$ since K' is the completion of K_0 ($=\{\Pi'(T)\xi' : T \in g\}$).

Condition (3) follows from:

$\Pi(T)\tau\ (S+N) = \Pi(T)(S+N_1) = TS+N_1 = \tau(TS+N) = \tau(\Pi'(T)(S+N))$ for T,S in $\mathcal{g}$. Condition (4) is a consequence of (3) and the fact that $\Pi(U)$ is a unitary operator for each U in G. ###

Proposition 3.11.: Let Π be a representation of $\mathcal{g}$ on K_1 such that :

(1) there is a unit vector $\eta \in K_1$ for which $K_1 = [\Pi(\mathcal{g})\eta]$.

(2) For each U in G, $\Pi(U)$ is an isometry.

Let g be the state on $\mathcal{g}$ defined by $g(T) = <\Pi(T)\eta,\eta>$ and let Π' be the representation of $\mathcal{g}$ associated with g. Then Π and Π' are unitarily equivalent.

Proof: a) If we were dealing with a self adjoint algebra instead of $\mathcal{g}$, we would have an equality of the form, $g(T^*T)=||\Pi(T)\eta||^2$, directly from our assumptions. We must obtain the corresponding information, in the present case, by other means.

In the first place, as noted earlier, g has a unique extension to a state of $\mathcal{B}(\mathcal{a},G)$. We denote by g, again, this state extension (so that $g(T^*T)$ refers to the value of this extension at T^*T, for T in $\mathcal{g}$).

Let T be $\sum_i U_iA_i$ where A_i in $\mathcal{a}$ and U_i in G, and assume that for any i, $U_i^*U_{i+1}$ is in G. Since $\Pi(U_i)$ is an isometry, we have $||\Pi(U_iA_i)\eta||^2=||\Pi(A_i)\eta||^2=g(A_i^*A_i)$ for each i; so that

$$<\Pi(T)\eta,\ \Pi(T)\eta>=\sum_{i,j}<\Pi(U_iA_i)\eta,\ \Pi(U_jA_j)\eta>$$
$$=\sum_i||\Pi(U_iA_i)\eta||^2+2\sum_{i>j}\mathrm{Re}<\Pi(U_iA_i)\eta,\ \Pi(U_jA_j)\eta>$$
$$=\sum_i g(A_iA_i)+2\sum_{i>j}\mathrm{RE}<\Pi(U_iA_i)\eta,\ \Pi(U_jA_j)\eta>.$$

For $i > j$, $U_j^*U_i \in G$, hence $<\Pi(U_iA_i)\eta\ ,\ \Pi(U_jA_j)\eta>$

$$= <\Pi(U_j)\Pi(U_j^*U_iA_i)\eta,\ \Pi(U_j)\Pi(A_j)\eta> =$$
$$= <\Pi(U_j^*U_iA_i)\eta,\ \Pi(A_j)\eta> =g(A_j^*U_j^*U_iA_i).$$

So $<\Pi(T)\eta,\ \Pi(T)\eta> =\sum_i g(A_i^*A_i)+2\sum_{i>j}\mathrm{Re}\ g(A_j^*U_j^*U_iA_i) = g(T^*T)$.

b) Define an operator $V:K' \longrightarrow K_1$ by $V\Pi'(T)\xi' = \Pi(T)\eta$, where K' is the representation space of Π', and $g(T) = <\Pi'(T)\xi',\xi'>$. Then V is well defined and isometric on a dense subspace of K' with range dense in K_1 since $||\Pi'(T)\xi'||^2=g(T^*T) = ||\Pi(T)\eta||^2$. Thus V can be extended to a unitary transformation of K' onto K_1 and $V\xi'=\eta$. We have $V\,\Pi'(T)V^*\Pi(S)\eta = V\Pi'(T)\Pi'(S)\xi' = \Pi(TS)\eta = \Pi(T)\,\Pi(S)\eta$; so that $V\,\Pi'(T)^*V = \Pi(T)$.

###

Definition: A state g of $\mathcal{B}(\mathcal{a},G)$ will be called G-invariant if for each T in $\mathcal{B}(\mathcal{a},G)$ and each U in G $g(U^*TU) = g(T)$.

Note that if g is a state and $g(U^*T^*TU) = g(T^*T)$ for each T in $g + g^*$ and each U in G, then g is G-invariant. In fact, since $\mathcal{B}(\mathcal{a},G)$ is the norm closure of $g + g^*$ and g is norm continuous; the equality, $g(U^*T^*TU) = g(T^*T)$, is valid for each T in $\mathcal{B}(\mathcal{a},G)$. But each positive operator in $\mathcal{B}(\mathcal{a},G)$ has the form T^*T for some T in $\mathcal{B}(\mathcal{a},G)$; and each operator in $\mathcal{B}(\mathcal{a},G)$ is a linear combination of at most four positive operators in $\mathcal{B}(\mathcal{a},G)$. Thus $g(U^*TU) = g(T)$ for each T in $\mathcal{B}(\mathcal{a},G)$ and each U in G.

Proposition 3.12.: Let g be a faithful G-invariant state of $\mathcal{B}(\mathcal{a},G)$ and let Π' be the representation of g associated with g. Then Π' is an isometry. i.e. $||\Pi'(T)|| = ||T||$.

Proof: Fix $\varepsilon > 0$. Let Π be the representation of $\mathcal{B}(\mathcal{a},G)$ associated with g, then $||\Pi(T)|| = ||T||$. Thus there is a unit vector x in K (the space on which Π acts) such that $||\Pi(T)x|| > ||T|| - \varepsilon$. With the notation of Proposition 3.10, there is an R in $g+g^*$ satisfying $g(R^*R) = 1$ and $||\Pi(T)(R+N_1)|| > ||T|| - \varepsilon$. Hence $(||T|| - \varepsilon)^2 < ||\Pi(TR)(I+N_1)||^2 = g(R^*T^*TR) = g(U^*R^*T^*TRU)$ for each U in G. Choose U such that $(S=)\ RU \in g$. Then, with T in g, $||\Pi'(T)(S+N||^2 = ||\Pi'(TS)(I+N)||^2 = ||\Pi'(TRU)(I+N)||^2$

$$= g(U^*R^*T^*TUR).$$

Thus $||\Pi'(T)(S+N)|| > ||T|| - \varepsilon$. Since ε is an arbitrary positive number and $g(S^*S) = g(U^*R^*RU) = g(R^*R) = 1$, we have that $||\Pi'(T)|| = ||T||$.

###

Boundedness of isomorphisms

Let $\mathcal{g}_1$, $\mathcal{g}_2$ be two irreducible triangular algebras generated by $\mathcal{a}_1, G_1$ and $\mathcal{a}_2$, G_2 respectively, and let Ψ be an isomorphism from $\mathcal{g}_1$ onto $\mathcal{g}_2$. We know that such a map is implemented by some (W,φ). In what follows we will find conditions for Ψ to be bounded. For this purpose we may assume that $W=I$.

Theorem 3.13.: If Ψ is an automorphism of $\mathcal{g}$ implemented by (I,φ) and g is a faithful state of $\mathcal{B}(\mathcal{a},G)$ satisfying, for A in $\mathcal{a}$ and U in $G\setminus\{I\}$: (1) $g(AU) = 0$

(2) $g(U^*A^*AU) = g(A^*A)$;

and there are positive constants N and M such that for each U in G $NI \leq |\varphi(U)| \leq MI$ then Ψ is bounded.

Proof: We note, first, that g is $\bar{G}$-invariant on $\mathcal{B}(\mathcal{a},G)$. Since g is G-invariant on the positive elements of $\mathcal{a}$ and such elements generate $\mathcal{a}$ (linearly); $g(U^*AU) = g(A)$ for each A in $\mathcal{a}$ and U in G. If $V \in \bar{G}\setminus G$ then $V^* \in G$ so that

$$g(A) = g(VV^*AVV^*) = g(V^*AV)$$

and $$g(AV)= \overline{g(V^*A^*)} = \overline{(V^*A^*VV^*)} = 0,$$

when $A \in \mathcal{a}$ (from (1), as $V^*A^*V \in \mathcal{a}$). Thus, if $T \in \mathcal{g}+\mathcal{g}^*$, $T = A_0+\sum_{j=1}^{n} A_jU_j$ where $A_j \in \mathcal{a}$ and $U_j \in \bar{G}\setminus\{I\}$, and with U in $\bar{G}$, $g(U^*TU)=g(U^*A_0U+\sum_{j=1}^{n}U^*A_jUU^*U_jU)=g(U^*A_0U)=g(A_0)=g(T)$. By norm continuity of g on $\mathcal{B}(\mathcal{a},G)$ it follows, now, that g is $\bar{G}$-invariant on $\mathcal{B}(\mathcal{a},G)$.

Let Π', K', and ξ' be the representation, the space, and the cyclic vector correspond to g. Define an operator V_0 on $\Pi'(\mathcal{g})\xi'$ by: $V_0\Pi'(T)\xi' = \Pi'(\Psi(T))\xi'$, where $T \in \mathcal{g}$. Then $||\Pi'(\Psi(T))\xi'||^2=g(\Psi(T)^*\Psi(T))$ and $||\Pi'(T)\xi'||^2 = g(T^*T)$.

Let T be $\sum AU_U$. Then, using the facts that $g(AU)=0$ for $U\neq I$ and $\Psi(A)=A$ for each A in $\mathcal{a}$ (recall that $W=I$ here) we have: $g(T^*T) = g(\sum U^*A_U^*A_UU) =$ $= \sum g(A_U^*A_U)$ and $g(\Psi(T)^*\Psi(T))=g(\sum U^*\varphi(U)^*A_U^*A_U\varphi(U)U) = \sum g(A_U^*|\varphi(U)|^2A_U)$.

By assumption (3): $N^2 g(T^*T) \leq g(\Psi(T)^*\Psi(T)) \leq M^2 g(T^*T)$. Thus $||V_0|| \leq M$ and V_0 can be extended to operator V on K' such that $||V|| \leq M$.

Since Ψ is an automorphism of $\mathcal{g}$ and $[\Pi'(\mathcal{g})\xi'] = K'$, V_0 has dense range; so that V has a bounded inverse and $||V^{-1}|| \leq N^{-1}$. As $V\xi' = \xi'$, we have $V\Pi'(T)V^{-1}\ \Pi'(S)\ \xi' = V\Pi'(T)\Pi'(\Psi^{-1}(S))\ \xi' = V\Pi'(T\Psi^{-1}(S))\xi'$ $= \Pi'(\Psi(T)S)\xi' = \Pi'(\Psi(T))\ \Pi'(S)\xi'$ for T and S in $\mathcal{g}$. Thus $V\Pi'(T)V^{-1} = \Pi'(\Psi(T))$; and $||V\Pi'(T)V^{-1}|| \leq ||V||\ ||\Pi'(T)||\ ||V^{-1}|| \leq N^{-1}M||T||$ hence $||\Pi'(\Psi(T))|| \leq N^{-1}M||T||$.

By Proposition 3.12, Π' is an isometry; so that $||\Psi(T)|| \leq N^{-1}M||T||$, and Ψ is bounded. ###

Remark: If $\bar{G}$ is amenable, g' is a faithful G-invariant state of $\mathcal{a}$, and g is $g'\cdot\Phi$, where Φ is the Arveson expectation of $\mathcal{B}(\mathcal{a},G)$ on $\mathcal{a}$ then g is a faithful $\bar{G}$-invariant state of $\mathcal{B}(\mathcal{a},G)$ (from the preceding theorem and the properties of Φ - see Lemma 2.1 and Proposition 2.7) for:

$g(U^*AU) = g'(\Phi(U^*AU)) = g'(U^*AU) = g'(A)$; and $g(AU) = g'(\Phi(AU)) = g'(0) = 0$,

for each A in $\mathcal{a}$ and U in $G\setminus\{I\}$. Thus g satisfies conditions (1) and (2) of the preceding theorem.

Corollary 3.14: If Ψ is an automorphism of $\mathcal{g}$ implemented by (I,φ) and $\mathcal{B}(\mathcal{a},G)$ has a faithful G-invariant state g such that $g(AU) = 0$ for each A in $\mathcal{a}$ and U in G then Ψ extends to a *-automorphism of $\mathcal{B}(\mathcal{a},G)$ if and only if $\varphi(U)$ is a unitary operator for each U in G.

Proof: If Ψ extends to a *-automorphism of $\mathcal{B}(\mathcal{a},G)$ then, of course, $\Psi(U)$ is unitary for each U in G. Now $\Psi(U) = \varphi(U)U$; so that $\varphi(U) = \Psi(U)U^*$, and $\varphi(U)$ is unitary for each U in G.

If $\varphi(U)$ is unitary for each U in G then condition (3) of the preceding theorem is satisfied (with 1 for N and M). From our assumptions, conditions (1) and (2) are also satisfied; and Ψ extends to a *-homomorphism on $\mathcal{B}(\mathcal{a},G)$. Since $\Psi(\mathcal{g}) = \mathcal{g}$; we have $\Psi(\mathcal{g}^*) = \mathcal{g}^*$ and Ψ maps $\mathcal{B}(\mathcal{a},G)$. onto $\mathcal{B}(\mathcal{a},G)$ (where we denote by Ψ, again, the extension of Ψ

to $\mathcal{B}(\mathfrak{a},G)$.

The automophism Ψ^{-1} of $\mathfrak{g}$ is implemented by (I,φ^*), where $\varphi^*(U) = \varphi(U)^*$; for $\Psi(U) = \varphi(U)U$ so that $U = \varphi(U)\Psi^{-1}(U)$ and $\Psi^{-1}(U) = \varphi(U) = \varphi(U)^*$. But $\varphi(U)^*$ is a unitary operator, for each U in G; so that, from what we have proved, to this point, Ψ^{-1} extends to a $*$-homomorphism (that we denote by Ψ^{-1}, again) of $\mathcal{B}(\mathfrak{a},G)$ onto itself. Now $\Psi^{-1}\cdot\Psi$ is the identity transformation on $\mathfrak{g}$ and, hence, on $\mathfrak{g}+\mathfrak{g}^*$. By (norm) continuity $\Psi^{-1}\cdot\Psi$ is the identity on $\mathcal{B}(\mathfrak{a},G)$; and Ψ is a $*$-automorphism of $\mathcal{B}(\mathfrak{a},G)$. ###

Isometries and anti-isomorphisms

A map $\Psi:\mathfrak{g} \longrightarrow B(K)$ is said to be an anti-isomorphism if it is linear one-to-one and **self adjoint** on $\mathfrak{a}$, and **satisfies**: $\Psi(TS)=\Psi(S)\Psi(T)$ $T,S \in \mathfrak{g}$. An anti-isomorphism Ψ will be said to be **implemented by** (W,φ) (where W is in $V(\mathfrak{a})$ and φ is a map from G to $\mathfrak{a}$) if for A in $\mathfrak{a}$, $\Psi(A)=W^*AW$ and, for U in G, $\Psi(U) = W^*U^*\varphi(U)W$.

Proposition 3.15.: Let $\Psi:\mathfrak{g}\longrightarrow\mathfrak{g}^*$ be an anti-isomorphism implemented by (I,φ). Assume that the conditions of Theorem 3.13 are fulfilled then Ψ is bounded.

Proof: With T in $\mathfrak{g}^*$, let $\tau(T)$ be T^* and let Ψ_1 be $\tau\cdot\Psi$. Then Ψ_1 is multiplicative and conjugate-linear. If $A \in \mathfrak{a}$, $U \in G$ then $\Psi_1(A) = A^*$, and $\Psi_1(U) = \varphi(U)^*U$.

The proof of Theorem 3.13 now applies to Ψ_1 with minor changes. In the present case, the operator V_0 is conjugate-linear; and $g(\Psi(T)^*\ \Psi(T)) = g(\sum U^*\varphi(U)A_U A_U^*\varphi(U)^*U) = \sum g(A_U A_U^*|\varphi(U)|^2)$. As in Theorem 3.13, we conclude that $V\Pi'(T)V^{-1} = \Pi'(\Psi_1(T))$; so that $||\Psi_1(T)|| = ||\Pi'(\Psi_1(T))|| \leq N^{-1}M||T||$. ###

Proposition 3.16.: The mapping Ψ defined by: $\Psi(AU) = U^*A$, for A in $\mathfrak{a}$, U in G, is an anti-isomorphism of $\mathfrak{g}$ onto $\mathfrak{g}^*$. If conditions (1), (2) of Theorem 3.13 are satisfied, then Ψ can be extended to a $*$-anti-isomorphism on $\mathcal{B}(\mathcal{a},G)$.

Proof.: From $\mathcal{a}$-invariance of G, Ψ is well defined (and linear). If $A,B \in \mathcal{a}$ and $U,V \in G$ then $\Psi(AU(BV)) = \Psi(AUBU^*UV) = V^*U^*AUBU^* = (V^*BU^*AUU^* = (V^*B)(U^*A)$

$$= \Psi(BV)\,\Psi(AU)\ .$$

By linearity, then, Ψ is an anti-isomorphism. Since Ψ satisfies the conditions of Proposition 3.15 (where $\varphi(U) = I$ for each U in G), Ψ is bounded (and $||\Psi|| = 1$).

Let $\Psi_0(S_1+S_2^*)$ be $\Psi(S_1) + \Psi(S_2)^*$, when $S_1,S_2 \in g$. Then Ψ_0 is a $*$-anti-automorphism of $g+g^*$. Note, too, that, with $\sum A_jU_j$ the (unique) representation of T in $g+g^*$ where $A_j \in \mathcal{a}$ and $\{U_j\}$ are distinct elements of $\bar{G}$, we have

$$\Psi_0(\Psi_0(T)) = \Psi_0(\sum U_j^*A_j) = \sum A_jU_j = T.$$

Moreover, we can choose U in G such that $TU \in g$. Then

$$\Psi(TU) = \Psi_0(TU) = \Psi_0(U)\,\Psi_0(T) = \Psi(U)\,\Psi_0(T) = U^*\Psi_0(T)\quad ;$$

so that

$$||\Psi_0(T)|| = ||\Psi(TU)|| \leq ||TU|| = ||T||.$$

Hence $||T|| = ||\Psi_0(\Psi_0(T))|| \leq ||\Psi_0(T)||$, and Ψ_0 is an isometry on $g+g^*$. It follows that Ψ_0 (and, hence, Ψ) extends to a $*$-anti-automorphism of $\mathcal{B}(\mathcal{a},G)$. ###

In [49, Theorem 6.4] E. Størmer proved the following: Let m and R be two unital C*-algebras and let $\Psi: m \longrightarrow R$ be a linear map satisfying:

(1) $\Psi(I) = I$.

(2) Ψ is positive.

(3) For any positive B in R, $\Psi^{-1}(B)$ contains a positive operator.

(4) $m_* \cap \operatorname{Ker}\Psi$ is linearily generated by the positive elements, where

$m_* = \{A \in m:\ A=A^*\}$

(5) Ψ is surjective.

Then, for any irreducible representation Π of R, the map $\Pi \cdot \Psi$ is a $*$-homomorphism or a $*$-anti-homomorphism.

Let $\mathcal{B}(\mathcal{a}_1,G_1)$ and $\mathcal{B}(\mathcal{a}_2,G_2)$ be the C*-algebras generated by the irreducible triangular algebras $\mathcal{g}_1$ and $\mathcal{g}_2$ respectively. Application of the result mentioned above gives:

Corollary 3.17.: Let Ψ be a linear isometry from $\mathcal{B}(\mathcal{a}_1,G_1)$ onto $\mathcal{B}(\mathcal{a}_2,G_2)$ which maps I into I. Then Ψ is either a *-isomorphism or a *-anti-isomorphism.

Proof.: Apply the result of Størmer to the identity representation. The conditions are satisfied because any isometry between C*-algebras is positive.

###

Similarily to what we showed for an isomorphism in Proposition 3.5 we can show that if Ψ is a *-anti-isomorphism from $\mathcal{g}+\mathcal{g}^*$ onto itself, mapping $\mathcal{a}$ onto $\mathcal{a}$, then it is implemented by some (W,φ). Also, we can prove Proposition 3.9 for a *-anti-isomorphism. Hence a *-anti-isomorphism on $\mathcal{B}(\mathcal{a},G)$ that maps $\mathcal{a}$ onto itself can be reduced to a *-anti-isomorphism on $\mathcal{g}+\mathcal{g}^*$. All this immediately implies the following:

Corollary 3.18.: Let Ψ be an isometry from $\mathcal{B}(\mathcal{a},G)$ onto itself that maps $\mathcal{a}$ onto $\mathcal{a}$ and I into I. Then there are a unitary operator $W \in V(\mathcal{a})$ and a map φ from G into $\mathcal{a}$ such that $\Psi(A) = W^*AW$ for A in $\mathcal{a}$, and for each U in G, $\Psi(U) = W^*\varphi(U)UW$ or $\Psi(U) = W^*U^*\varphi(U)W$. ###

The order is preserved by isomorphisms

In chapter II we defined an ordering on the group $M(\mathcal{g})$, with $N(\mathcal{g})$ its positive cone. We will see that an isomorphism preserves this ordering. Let $\mathcal{g}_1$, $\mathcal{g}_2$ be irreducible triangular algebras generated by $\mathcal{a}_1$, G_1 and $\mathcal{a}_2,G_2$ respectively. Throughout this section, Ψ will be an isomorphism from $\mathcal{g}_1$ onto $\mathcal{g}_2$ implemented by (W,φ).
We will denote by Φ_1 and Φ_2 the Arveson expectations onto $\mathcal{a}_1$ and $\mathcal{a}_2$ respectively.

Proposition 3.19.: The isomorphism Ψ maps $N(\mathcal{g}_1)$ onto $N(\mathcal{g}_2)$. If Ψ can be extended to a $*$-isomorphism from $\mathcal{G}_1+\mathcal{G}_1^*$ onto $\mathcal{G}_2+\mathcal{G}_2^*$ then this extension maps $M(\mathcal{G}_1)$ onto $M(\mathcal{G}_2)$.

Proof.: An element in $N(\mathcal{G}_1)$ has the form AV, where A is invertible in $\mathcal{a}_1$, $V \in V(\mathcal{a}_1) \cap \mathcal{G}_1$, and (from Lemma 2.11) we can write $V = \sum F_iU_i$ where U_i are in G_1, and F_i are projections in $\mathcal{a}_1$ with sum I. Since $F_iF_j = 0$, $i \neq j$, $F_iV = F_iU_i$ and: $\Psi(AV)=W^*A(\sum\varphi(U_i)F_iU_i)W=W^*A(\sum\varphi(U_i)F_i)VW=BW^*VW$, where $B=W^*A(\sum\varphi(U_i)F_i)W$. Moreover, B is an invertible element of $\mathcal{a}_2$ because A and $\varphi(U_i)$ are invertible (cf. Proposition 3.5) and $\sum F_i = I$.

Since $\Psi|\mathcal{a}$ is adjoint preserving, $\Psi(\mathcal{a}_1) \subseteq \mathcal{a}_2$. By definition of "implementation", $W^*\mathcal{G}_1W = \mathcal{G}_2$, so that $W\mathcal{a}_2W^* \subseteq \mathcal{G}_1$ and, as $W\mathcal{a}_2W^*$ is self adjoint, $W\mathcal{a}_2W^* \subseteq \mathcal{a}_1$. Thus $\Psi(\mathcal{a}_1) = W^*\mathcal{a}_1W = \mathcal{a}_2$.

For any C in $\mathcal{a}_2$, WCW^* is in $\mathcal{a}_1$, hence V^*WCW^*V is in $\mathcal{a}_1$ and $W^*V^*WCW^*VW$ is in $\mathcal{a}_2$, hence W^*VW is in $V(\mathcal{a}_2) \cap \mathcal{G}_2$. Thus Ψ maps $N(\mathcal{G}_1)$ into $N(\mathcal{G}_2)$. Now Ψ^{-1} is an isomorphism of $\mathcal{G}_2$ onto $\mathcal{G}_1$; and we have just noted that Ψ restricted to $\mathcal{a}_1$ is a $*$-isomorphism of $\mathcal{a}_1$ onto $\mathcal{a}_2$ so that Ψ^{-1} restricted to $\mathcal{a}_2$ is a $*$-isomorphism of $\mathcal{a}_2$ onto $\mathcal{a}_1$. From Proposition 3.2, Ψ^{-1} is implemented by (W^*,φ'); and, by what we have proved, Ψ^{-1} maps $N(\mathcal{G}_2)$ into $N(\mathcal{G}_1)$. Hence Ψ maps $N(\mathcal{G}_1)$ onto $N(\mathcal{G}_2)$.

For the second assertion of this proposition, note that if Ψ extends to a $*$-isomorphism Ψ' of $\mathcal{G}_1+\mathcal{G}_1^*$ onto $\mathcal{G}_2+\mathcal{G}_2^*$ and $U \in V(\mathcal{a}_1) \cap (\mathcal{G}_1+\mathcal{G}_1^*)$ then $\Psi(U) \in V(\mathcal{a}_2) \cap (\mathcal{G}_2+\mathcal{G}_2^*)$; for each operator in $\mathcal{a}_2$ has the form, $\Psi(A)$, with A in $\mathcal{a}_1$, and

$$\Psi(U)^*\Psi(A)\Psi(U) = \Psi(U^*AU) \in \mathcal{a}_2.$$

Thus, with AU in $M(\mathcal{G}_1)$, $\Psi(AU) \in M(\mathcal{G}_2)$. It follows (applying the same considerations to Ψ^{-1}) that Ψ maps $M(\mathcal{G}_1)$ onto $M(\mathcal{G}_2)$. ###

In connection with the preceding proposition, recall (from Proposition 3.1) that Ψ extends to a $*$-isomorphism of $\mathcal{G}_1+\mathcal{G}_1^*$ if and only if $\Psi(U)$ is unitary for each U in G_1. As a consequence, we have:

Corollary 3.20.: If Ψ is an isomorphism of $\mathcal{G}_1$ onto $\mathcal{G}_2$ and S,T in

$N(\mathcal{g}_1)$ satisfy $S > T$ then $\Psi(S) > \Psi(T)$.

If Ψ is an isomorphism (not necessarily *-) of $\mathcal{g}_1+\mathcal{g}_1^*$ onto $\mathcal{g}_2+\mathcal{g}_2^*$ that maps $\mathcal{g}_1$ onto $\mathcal{g}_2$ and S,T in $M(\mathcal{g}_1)$ satisfy $S > T$ then $\Psi(S) > \Psi(T)$.

Lemma 3.21.: For any T in $\mathcal{g}_1$ $\Phi_2\cdot\Psi(T) = \Psi\cdot\Phi_1(T)$. If Ψ can be extended to a *-isomorphism of $\mathcal{g}_1+\mathcal{g}_1^*$ then the equality holds for any T in $\mathcal{g}_1+\mathcal{g}_1^*$ and if it can be further extended to a *-isomorphism of $\mathcal{B}(\mathcal{a},G)$ then it will hold for any T in $\mathcal{B}(\mathcal{a},G)$.

Proof: It is enough to prove the statement concerning $\mathcal{g}_1$. The statements about $\mathcal{g}_1+\mathcal{g}_1^*$ and $\mathcal{B}(\mathcal{a},G)$ will follow from self-adjointness and norm continuity of $\Phi_2\cdot\Psi$ and $\Psi\cdot\Phi_1$.

Let $T = \sum A_iU_i$, where $A_i \in \mathcal{a}$, $U_i \in G_1$, and $U_0 = I$. Then $\Psi(T) = W^*(\sum A_i\varphi(U_i)U_i)W = \sum W^*A_i\varphi(U_i)W(W^*U_iW)$. If $\Phi_2(W^*U_iW) \neq 0$ there is a non zero projection P in $\mathcal{a}_2$ such that $E\Phi_2(W^*U_iW) = 0$ for each non zero subprojection E of P in $\mathcal{a}_2$. But then $0 \neq E\Phi_2(W^*U_iW) = E\Phi_2(W^*U_iW)E = \Phi_2(EW^*U_iWE) = \Phi_2(W^*WEW^*U_iWEW^*W)$, for such subprojection E. In particular $0 \neq FU_iF$ for each subprojection $F(=WEW^*)$, in $\mathcal{a}_1$, of Q $(=WPW^*)$. Hence $0 \neq FU_iFU_i^*$ for each such F - contradicting the assumption that U_i acts freely on $\mathcal{a}_1$. Thus $\Phi_2(W^*U_iW) = 0$ for each i; and $\Phi_2(\Psi(T)) = W^*A_0W$. On the other hand, $\Phi_1(T) = A_0$; so that

$$\Psi(\Phi_1)(T)) = W^*A_0W = \Phi_2(\Psi(T)).$$ ###

Corollary 3.22.: If the isomorphism Ψ of $\mathcal{g}_1$ onto $\mathcal{g}_2$ extends to a *-isomorphism of $\mathcal{B}(\mathcal{a}_1,G_1)$ onto $\mathcal{B}(\mathcal{a}_2,G_2)$ then Φ_2 is faithful if and only if Φ_1 is faithful. If, in addition, G_1 and G_2 are unitary semigroups arising from measure preserving transformations of a finite measure space, then G_1 is amenable if and only if G_2 is amenable.

Proof: Since $\Phi_2\cdot\Psi' = \Psi'\cdot\Phi_1$, where Ψ' is the extension of Ψ to a *-isomorphism of $\mathcal{B}(\mathcal{a}_1,G_1)$ onto $\mathcal{B}(\mathcal{a}_2,G_2)$, from the preceding lemma; Φ_2 is faithful if and only if Φ_1 is faithful. From Proposition 2.7, the discussion following it, and Proposition 2.8, with G_1 and G_2 semigroups of measure preserving transformations, G_1 is amenable if and only if Φ_1 is faithful and G_2 is amenable if and only if Φ_2 is faithful. Thus, in this case, G_1 is amenable if and only if G_2 is amenable. ###

Corollary 3.23.: If Ψ is an isomorphism and $N(\mathcal{g}_1)$ has a smallest element T, then $\Psi(T)$ is a smallest element in $N(\mathcal{g}_2)$. If R is another smallest element in $N(\mathcal{g}_2)$ then $\Psi(T)$ is in $\mathcal{a}_2R$.

Proof: By Lemma 3.21 we see that $\Phi_1(T) = 0$, for T in $\mathcal{g}_1$, if and only if $\Phi_2(\Psi(T)) = 0$. This, with Proposition 3.19, Corollary 3.20, and the definition of a smallest element, completes the proof. ###

Similarily we can get, from the results of Chapter II:

Corollary 3.24.: If Ψ is an isomorphism:

(1) G_1 is integrally closed if and only if G_2 is.

(2) G_1 has a minimal element if and only if G_2 has one.

(3) G_1 is singly generated if and only if G_2 is.

Proof:

(1) is a result of Proposition 2.27 and the fact that the isomorphism is multiplicative and order preserving.

(2) is a result of Lemma 2.24 and Corollary 3.23.

(3) is a result of Corollary 3.21, Proposition 2.27, and Corollary 2.28.

###

From the preceding we learn that an algebra generated by a singly generated semigroup cannot be isomorphic to an algebra whose semigourp is not singly generated. This gives us many examples of non isomorphic irreducible triangular algebras.

In the following example we will construct two non isomorphic irreducible triangular algebras. The semigroups that generate them are each generated by two operators and the two algebras generate the same self adjoint algebra.

Example 3.25.: Let H be the Hilbert space $L^2(\mathbb{T})$, and $\mathcal{a}$ be the maximal abelian algebra of multiplication by functions in $L^\infty(\mathbb{T})$. Let r and s be two real irrational numbers satisfying $(\mathbb{Z}r+\mathbb{Z}s)\cap\mathbb{Z}=\{0\}$ (e.g. $r=\sqrt{2}$, $s=\sqrt{3}$).

Let U and V be the following operators:

$$Uf(e^{2\Pi it}) = f(e^{2\Pi i(t-r)})$$

$$Vf(e^{2\Pi it}) = f(e^{2\Pi i(t-s)}).$$

Define two semigroups:

$G_1 = \{V^kU^n : k,n\in\mathbb{Z}, k>0 \text{ or } k=0 \text{ and } n\geq 0\}.$

$G_2 = \{V^kU^n : k,n\in\mathbb{Z}, n+k\sqrt{2}\geq 0\}.$

For each pair (k,n) in $\mathbb{Z}\times\mathbb{Z}\setminus\{(0,0)\}$, the operator V^kU^n corresponds to the translation by $ks+nr$; this number is irrational (as $(\mathbb{Z}s+\mathbb{Z}r)\cap\mathbb{Z}=\{0\}$) hence V^kU^n acts freely and ergodically on $\mathcal{a}$ (see Example 1.8). Thus G_1 and G_2 each acts freely and ergodically on $\mathcal{a}$.

Since $G_1^*\cup G_1 = \{V^kU^n: (k,n)\in\mathbb{Z}\times\mathbb{Z}\}$ and U commutes with V, $G_1^*\cup G_1$ $(=\bar{G}_1)$ is a group homomorphic to the additive group $\mathbb{Z}\times\mathbb{Z}$ via the homomorphism $\tau(k,n) = V^kU^n$.

As $(\mathbb{Z}r+\mathbb{Z}s)\cap\mathbb{Z} = \{0\}$, $V^kU^n = I$ only if $n=k=0$. Hence τ is an isomorphism.

Define the lexicographic ordering on $\mathbb{Z}\times\mathbb{Z}$ by: $(k,n)\geq(1,m)$ if $k>1$ or $k=1$ and $n\geq m$.

The positive cone of this ordering is the set p of all pairs (k,n) such that $k>0$ or $k=0$ and $n\geq 0$. From the definition of G_1, τ maps p onto G_1. Since $p\cap(-p) = \{(0,0)\}$, $G_1\cap G_1^{-1} = \{I\}$, hence G_1 is an ordered semigroup, order-isomorphic to p. The pair $(0,1)$ is minimal in $p\setminus\{(0,0)\}$ hence U $(=\tau(0,1))$ is a minimal element in G_1.

Thus G_1 is an ordered semigroup, with a minimal element, that acts freely and ergodically on $\mathcal{a}$. It gives rise to an irreducible triangular algebra $\mathcal{g}_1$ (by Corollary 1.5).

For G_2, let Λ be the set $\mathbb{Z}+\sqrt{2}\mathbb{Z}$ $(\subseteq \mathbb{R})$ and define a map $\eta:\Lambda \longrightarrow G_2 \cup G_2^{-1} (=\bar{G}_2)$ by $\eta(n+k\sqrt{2}) = V^k U^n$. As $\mathbb{Z}\cap\mathbb{Z}\sqrt{2} = \{0\}$, η is well defined and is a homomorphism from Λ (an additive group) onto $\bar{G}_2$.
If $n+k\sqrt{2} \neq 0$ then $(n,k) \neq (0,0)$, hence $V^k U^n \neq I$; so that η is an isomorphism.
If we let Λ have the ordering induced from $\mathbb{R}$ then, by the definition of G_2, η preserves the ordering and maps the set of non negative numbers in Λ (to be denoted by Λ_+) onto G_2. Since $\Lambda_+\cap(-\Lambda_+) = \{0\}$, $G_2 \cap G_2^{-1} = \{I\}$; so that G_2 is an ordered semigroup. To see that G_2 has no minimal element, let $\{r_k\}$ be a sequence of rational numbers such that $r_k \longrightarrow \sqrt{2}$ and $r_k > \sqrt{2}$ for each k. For each k, let m_k, n_k be positive integers such that $r_k = m_k/n_k$, then $m_k/n_k - \sqrt{2} \longrightarrow 0$ and $m_k/n_k - \sqrt{2} > 0$, hence $m_k - n_k\sqrt{2} \longrightarrow 0$ and $m_k - n_k\sqrt{2} \in \Lambda_+\setminus\{0\}$.
Thus $\Lambda_+\setminus\{0\}$ has no minimal number, hence G_2 has no minimal element.
The semigroup G_2 acts freely and ergodically on $\mathcal{a}$ and gives rise to an irreducible triangular algebra $\mathcal{g}_2$.

Since G_1 has a minimal element and G_2 has no such element, Corollary 3.24 implies that $\mathcal{g}_1$ is not isomorphic to $\mathcal{g}_2$.

CHAPTER IV: AUTOMORPHISMS AND DERIVATIONS

In this chapter we will study groups of $*$-automorphisms on $\mathcal{B}(\mathcal{a},G)$ (where $\mathcal{a}$ is a maximal abelian algebra and G is a semigroup of unitary operators acting freely and ergodically on $\mathcal{a}$) that map $\mathcal{a}$ onto itself, and their relationship to derivations of $\mathcal{B}(\mathcal{a},G)$ (or $g+g^*$) that vanish on $\mathcal{a}$.

We will use the notions of spectrum and spectral spaces, and the theory introduced by Arveson in [5].

Let $\mathrm{Aut}(\mathcal{B},\mathcal{a})$ denote the group of $*$-automorphisms Ψ on $\mathcal{B}(\mathcal{a},G)$ satisfying $\Psi(A) = A$ for each A in $\mathcal{a}$. By Corollary 3.8 (2) every Ψ in $\mathrm{Aut}(\mathcal{B},\mathcal{a})$ is implemented by some (W,φ'). Since $A = \Psi(A) = W^*AW$, $W \in \mathcal{a}' = \mathcal{a}$ and $\Psi(U)$ $(=W^*\varphi'(U)UW = W^*\varphi'(U)UWU^*)U)$ is in $\mathcal{a}U$ for each U in $\bar{G}$. Letting $\varphi(U)$ be $W^*\varphi'(U)UWU^*$ we conclude that Ψ is implemented by (I,φ).

Throughout the chapter Λ will denote a locally compact abelian group with a Haar measure m. We denote by $L^1(\Lambda)$ the (classes of) m-integrable functions on Λ. With the norm $||f||_1 = \int_\Lambda |f(t) \, dm(t)$ the space $L^1(\Lambda)$ is a Banach space.

Let $M(\Lambda)$ denote the Banach space of all complex finite regular Borel measures on Λ with the total variation as the norm. For each f in $L^1(\Lambda)$ the measure $f(t)dm(t)$ is in $M(\Lambda)$ and its total variation equals $||f||_1$.

For a Banach space X let $X^\#$ be the dual space; i.e. the space of all linear bounded functionals on X. With the norm $||g|| = \mathrm{Sup}\,\{|g(x)| : x \in X, \ ||x|| \leq 1\}$ $X^\#$ is a Banach space.

The following result is a special case of Proposition 1.2 of [5] and will be stated here without a proof.

Proposition 4.1.: Let X be a Banach space and let β be a map from the group Λ into X such that $\mathrm{Sup}\{||\beta(t)|| : t \in \Lambda\} < \infty$. Assume that for each

g in $X^{\#}$ the map $t \longrightarrow g(\beta(t))$ is continuous, then, for each μ in $M(\Lambda)$, there is a vector x in X such that for every g in $X^{\#}$ $g(x) = \int_\Lambda g(\beta(t))d\mu(t)$.

We will employ the usual notation for this integral:

$$x = \int_\Lambda \beta(t)d\mu(t).$$

The vector x, above, is unique with this property since the functionals in $X^{\#}$ separate vectors in X.

Let α be a homomorphism from Λ into $\mathrm{Aut}(\mathcal{B},\mathcal{a})$ such that for every T in $\mathcal{B}(\mathcal{a},G)$ the map $t \longrightarrow \alpha_t(T)$ is norm-continuous. As noted above, for every t in Λ, α_t is implemented by (I,φ_t) for some map φ_t from $\bar{G}$ into $\mathcal{a}$.

If U is in $\bar{G}$ and t,s are in Λ then

$$\begin{aligned}\varphi_{t+s}(U) &= \alpha_{t+s}(U)U^* = \alpha_t(\alpha_s(U))U^* = \alpha_t(\varphi_s(U)U)U^* = \alpha_t(\varphi_s(U))\ \alpha_t(U)U^* \\ &= \varphi_s(U)\ \varphi_t(U).\end{aligned}$$

Thus the map $t \longrightarrow \varphi_t(U)$ is a homomorphism for each U in $\bar{G}$. we will denote this homomorphism by φ_U.

Corollary 4.2.:

(1) For every μ in $M(\Lambda)$ and T in $\mathcal{B}(\mathcal{a},G)$ there is a unique bounded operator $\alpha_\mu(T)$ in $\mathcal{B}(\mathcal{a},G)$ such that for each g in $\mathcal{B}(\mathcal{a},G)^{\#}$

$$g(\alpha_\mu(T)) = \int_\Lambda g(\alpha_t(T))d\mu(t).$$

(2) For every μ in $M(\Lambda)$ and U in $\bar{G}$ there is a unique bounded operator $\varphi_\mu(U)$ in $\mathcal{a}$ such that for each g in $\mathcal{a}^{\#}$

$$g(\varphi_\mu(U)) = \int_\Lambda g(\varphi_t(U))d\mu(t).$$

Proof: For (1) apply Proposition 4.1 with $\mathcal{B}(\mathcal{a},G)$ in place of X and $\alpha_t(T)$ in place of $\beta(t)$. For (2) apply the same proposition with $\mathcal{a}$ in place of X and φ_U in place of β (since $\varphi_U(t) = \alpha_t(U)U^*$; φ_U is norm-continuous). ###

We will write $\alpha_\mu(T) = \int_\Lambda \alpha_t(T)d\mu(t)$

and $\varphi_\mu(U) = \int_\Lambda \varphi_t(U)d\mu(t)$.

We will write $\hat{\Lambda}$ for the dual group of Λ ; i.e. $\hat{\Lambda}$ is the group of all characters of Λ.
Equipped with the topology of uniform convergence on compact subsets of Λ, the dual group $\hat{\Lambda}$ is a locally compact abelian group.

For f in $L^1(\Lambda)$ we write

$\hat{f}(\tau) = \int_\Lambda \tau(t)f(t)dm(t)$ for $\tau \in \hat{\Lambda}$ (Fourier transform).

Let $K(\Lambda)$ denote the dense ideal of $L^1(\Lambda)$ consisting of the functions f whose Fourier transform has a compact support (in $\hat{\Lambda}$). We will write Supp f for the support of $\hat{f}$ (in $\hat{\Lambda}$).

By the w-topology on $\mathcal{B}(\mathcal{a},G)$ we mean the weak topology induced by the dual of $\mathcal{B}(\mathcal{a},G)$ i.e. T_n tend to T in this topology if and only if $g(T_n)$ tend to $g(T)$ for each linear bounded functional g on $\mathcal{B}(\mathcal{a},G)$.

Let Ω be an open set in $\hat{\Lambda}$. Denote by $K(\Lambda,\Omega)$ the set of functions f in $K(\Lambda)$ such that the support of $\hat{f}$ is contained in Ω, and define $R^\alpha(\Omega)$ to be the w-closed (equivalently norm-closed) linear space spanned by the set of all operators of the form $\alpha_f(T)$ for T in $\mathcal{B}(\mathcal{a},G)$ and f in $K(\Lambda,\Omega)$.

Lemma 4.3.: For every μ in $M(\Lambda)$, U in $\bar{G}$, and A in $\mathcal{a}$

$\alpha_\mu(AU) = A\varphi_\mu(U)U$.

Proof.: Fix U in $\bar{G}$, A in $\mathcal{a}$, and μ in $M(\Lambda)$. For g in $\mathcal{B}^\#$:

$g(\alpha_\mu(AU)) = \int_\Lambda g(\alpha_t(AU))d\mu(t) = \int_\Lambda g(A\varphi_t(U)U)d\mu(t)$.

Let g_0, in the dual of $\mathcal{a}$, be defined by $g_0(B) = g(BAU)$. As g is linear, g_0 is linear and $|g_0(B)| = |g(BAU)| \leq ||g||\ ||B||\ ||A||$; so that g_0 is in $\mathcal{a}^\#$.

By the definition of $\varphi_\mu(U)$ we now have:

$g(\alpha_\mu(AU)) = \int_\Lambda g_0(\varphi_t(U))d\mu(t) = g_0(\varphi_\mu(U)) = g(A\ \varphi_\mu(U)U)$.

Since this holds for each g in the dual of $\mathcal{B}(\mathcal{a},G)$ we obtain

$\alpha_\mu(AU) = A\varphi_\mu(U)U$. ###

Lemma 4.4.: For each open subset Ω of $\hat{\Lambda}$, $R^\alpha(\Omega)$ is the w-closed linear space spanned by the operators of the form $A\alpha_f(U)U$ where f is in $K(\Lambda,\Omega)$. A in $\mathcal{a}$, and U is in $\bar{G}$.

Proof.: Since $\alpha_f(AU) = A\varphi_f(U)U$, the operators of this form are in $R^\alpha(\Lambda)$, for f in $K(\Lambda,\Omega)$, A in $\mathcal{a}$ and U in $\bar{G}$. Denote the w-closed linear space they span by R'. Hence $R' \subseteq R^\alpha(\Lambda)$ and it will suffice to prove that $R^\alpha(\Lambda) \subseteq R'$.

Any T in $g+g^*$ can be expressed as $\sum_{i=1}^n A_iU_i$ where A_1 are in $\mathcal{a}$ and U_i are in $\bar{G}$. Thus $\alpha_f(T) = \sum_{i=1}^n A_i\varphi_f(U_i)U_i$ hence $\alpha_f(T)$ is in R'. Let T be in $\mathcal{B}(\mathcal{a},G)$. Then there are operators T_n in $g+g^*$ tending in norm to T. Now,, $\alpha_f(T)$ is an element of $\mathcal{B}(\mathcal{a},G)$ such that

$g(\alpha_f(T)) = \int_\Lambda g(\alpha_t(T))f(t)dm(t)$ for each g in $\mathcal{B}(\mathcal{a},G)^\#$. If $||g|| \leq 1$ then $|g(\alpha_f(T))| = |\int_\Lambda g(\alpha_t(T))f(t)dm(t)| \leq \int_\Lambda |g(\alpha_t(T))||f(t)\ dm(t) \leq$

$||T||\ ||f||_1$.

Since this holds for each g in the unit ball of $\mathcal{B}^\#$,

$$||\alpha_f(T)|| \leq ||T||\ ||f||_1.$$

From the defining property of α_μ, for μ in $M(\Lambda)$, α_μ is a linear mapping of $\mathcal{B}(\mathcal{a},G)$ into itself. The preceding inequality tells us that

$||\alpha_f|| \leq ||f||_1$. Hence $\alpha_f(T_n)$ tends to $\alpha_f(T)$ in norm and therefore also in the w-topology. Thus, for each T in $\mathcal{B}(\mathcal{a},G)$ and each f in $K(\Lambda,\Omega)$, $\alpha_f(T)$ is in R'. Since those operators span $R^\alpha(\Omega)$, $R^\alpha(\Omega)$ is contained in R'. ###

For a closed subset Z of $\hat{\Lambda}$ we define $M^\alpha(Z)$ to be the set of all operators T in $\mathcal{B}(\mathcal{a},G)$ such that $\alpha_f(T) = 0$ for each f in $K(\Lambda,\hat{\Lambda}\backslash Z)$.

Let T in $g+g^*$ be $\sum A_UU$, then T is in $M^\alpha(Z)$ if and only if $\sum\varphi_f(U)A_UU = 0$ for all f in $K(\Lambda,\hat{\Lambda}\backslash Z)$, Hence T is in $M^\alpha(Z)$ if and only if for every f in $K(\Lambda,\hat{\Lambda}\backslash Z)$ $\varphi_f(U)A_U = 0$. We, therefore, conclude:

Lemma 4.5: $M^\alpha(Z) = \mathcal{B}(\mathcal{a},G)$ if and only if for each f in $K(\Lambda,\hat{\Lambda}\backslash Z)$ and each U in $\bar{G}$, $\varphi_f(U) = 0$.

Proof: If $M^\alpha(Z) = \mathcal{B}(\mathcal{a},G)$ then, for each f in $K(\Lambda,\hat{\Lambda}\backslash Z)$ and each $T(=\sum A_UU)$

in $\mathcal{g}+\mathcal{g}^*$, $\varphi_f(U)A_U = 0$. Hence $\varphi_f(U) = 0$ for each U in G.

For the other direction, note that $M^\alpha(Z)$ is closed in the norm topology hence it will suffice to show that each T in $\mathcal{g}+\mathcal{g}^*$ is in $M^\alpha(Z)$. The discussion preceding this lemma completes the proof. ###

The <u>spectrum</u> of α is defined as the smallest closee set Z in Λ such that $M^\alpha(Z) = \mathcal{B}(\mathcal{a},G)$ and is denoted by $sp(\alpha)$. For the spectrum of an operator T we will use the notation $\sigma(T)$.

Let Λ be $\mathbb{R}$ and fix U in $\bar{G}$, then $t \longrightarrow \varphi_t(U)$ is a norm continuous one parameter unitary group. Hence there is a self adjoint operator $C(U)$ in $\mathcal{a}$ such that $\varphi_t(U) = \exp(itC(U))$ by [11, Theorem VII, 1.2]).

<u>Theorem 4.6.:</u> Let $t \longrightarrow \alpha_t$ be a homomorphism from $\mathbb{R}$ into $\mathrm{Aut}(\mathcal{B},\mathcal{a})$ implemented by (I,φ_t), such that for each T in $\mathcal{B}(\mathcal{a},G)$ the map $t \longrightarrow \alpha_t(T)$ is norm continuous. For U in $\bar{G}$, let $C(U)$ be the operator in $\mathcal{a}$ such that $\exp(itC(U)) = \varphi_t(U)$, then $sp(\alpha) = \overline{\bigcup_{U\in\bar{G}} \sigma(C(U))}$ where $\bar{Y}$ denotes the closure of the set Y.

<u>Proof.</u>: a) Fix U in $\bar{G}$. If t' is in $\sigma(C(U))$ but not in $sp(\alpha)$ then there is f in $K(\mathbb{R})$ with $\hat{f}(t)=1$ and $\mathrm{Supp}\ f \subseteq \mathbb{R}\setminus sp(\alpha)$. Since $M^\alpha(sp(\alpha))$ is $\mathcal{B}(\mathcal{a},G)$, Lemma 4.5 implies that $\varphi_f(U)=0$; hence $0=\varphi_f(U)=\int \varphi_t(U)f(t)dm(t)=\int_{\mathbb{R}} \exp(it\ (C(U)))f(t)dm(t)$. Let τ be a pure state of $\mathcal{a}$ and let s be $\tau(C(U))$ then:

$$0 = \tau(\varphi_f(U)) = \int_{\mathbb{R}} \exp(it\ \tau(C(U)))f(t)dm(t)$$

$$= \int_{\mathbb{R}} \exp(its)f(t)dm(t) = \hat{f}(s).$$

But, since t' is in $\sigma(C(U))$, $t' = \tau(C(U))$ for some pure state τ, therefore $\hat{f}(t') = 0$. This contradicts the choice of t' and the contradiction proves that $\sigma(C(U))$ is contained in $sp(\alpha)$.

Since U is an arbitrary member of $\bar{G}$ and since $sp(\alpha)$ is closed, $\overline{\bigcup_{U\in\bar{G}} \sigma(C(U))} \subseteq sp(\alpha)$.

b) Let f be in $K(\mathbb{R})$ with $\mathrm{Supp}\ \hat{f}$ contained in $\overline{\bigcup_{U\in\bar{G}} \sigma(C(U))}$, then for each U in $\bar{G}$. $\mathrm{Supp}\ \hat{f} \subseteq \mathbb{R}\setminus\overline{\sigma(C(U))}$.

Fix U in $\bar{G}$, then for each pure state τ of $\mathcal{a}$ $\tau(\varphi_f(U)) = \hat{f}(\tau(C(U))) = 0$, thus $\varphi_f(U) = 0$.

Since this holds for each U in $\bar{G}$, it implies, by Lemma 4.5, that $M^{\alpha}(\overline{\cup_{U \in \bar{G}} \sigma(C(U))}) = \mathcal{B}(\mathcal{a},G)$. Hence $sp(\alpha) \subseteq \overline{\cup_{U \in \bar{G}} \sigma(C(U))}$ and the proof is complete. ###

Proposition 4.7.: Let α_1 be a $*$-automorphism in $Aut(\mathcal{B},\mathcal{a})$ and let α be the homomorphism from $\mathbb{Z}$ into $Aut(\mathcal{B},\mathcal{a})$ that maps n into α_1^n. If α_1 is implemented by (I,φ) then $sp(\alpha) = \overline{\cup_{U \in \bar{G}} \sigma(\varphi(U))}$.

Proof: The proof is similar to the proof of the preceding theorem.

a) If t is in $\mathbb{T} \setminus sp(\alpha)$ there is some f in $K(\mathbb{Z})$ with $\hat{f}(t) = 1$ and $\text{Supp}\,\hat{f} \subseteq \mathbb{T} \setminus sp(\alpha)$. By Lemma 4.5, applied to $Z = sp(\alpha)$, $\varphi_f(U) = 0$ hence $0 = \varphi_f(U) = \sum_{n=-\infty}^{\infty} \varphi_n(U) f(n)$.

Since $n \longrightarrow \varphi_n(U)$ is a homomorphism (see the discussion preceding Corollary 4.2) and $\varphi(U) = \varphi_1(U)$, $\varphi_n(U) = \varphi(U)^n$ for each n in $\mathbb{Z}$ and each U in $\bar{G}$. Let τ be a pure state of $\mathcal{a}$, and let s be $\tau(\varphi(U))$, then

$$0 = \tau(\varphi_f(U)) = \sum \tau(\varphi(U))^n f(n) = \sum s^n f(n) = \hat{f}(s).$$

But, if t is in the spectrum of $\varphi(U)$ $(\subseteq \mathbb{T})$, t equals $\tau(\varphi(U))$ for some pure state τ of $\mathcal{a}$; and $\hat{f}(t) = 0$. This contradicts the choice of f and proves that, for each U in $\bar{G}$, $\sigma(\varphi(U)) \subseteq sp(\alpha)$.

Since $sp(\alpha)$ is closed, $\overline{\cup_{U \in \bar{G}} \sigma(\varphi(U))} \subseteq sp(\alpha)$.

b) Let f be in $K(\mathbb{Z})$ with $\text{Supp}\,\hat{f}$ contained in $\mathbb{T} \setminus \overline{\cup_{U \in \bar{G}} \sigma(\varphi(U))}$, then for each U in $\bar{G}$ $\text{Supp}\,\hat{f} \subseteq \mathbb{T} \setminus \sigma(\varphi(U))$. For such f in $K(\mathbb{Z})$, U in $\bar{G}$, and a pure state τ of $\mathcal{a}$ $\tau(\varphi_f(U)) = \hat{f}(\tau(\varphi(U))) = 0$, thus $\varphi_f(U) = 0$. This, by Lemma 4.5, implies that $M^{\alpha}(\overline{\cup_{U \in \bar{G}} \sigma(\varphi(U))})$ is $\mathcal{B}(\mathcal{a},G)$ hence $sp(\alpha) \subseteq \overline{\cup_{U \in \bar{G}} \sigma(\varphi(U))}$.. This completes the proof. ###

We now state, without proof, two general results (Corollary 8.1.11 and Theorem 8.1.12 from [42]).

Corollary 4.8.: Let β be an invertible isometry on a Banach space X, and consider the representation $n \longrightarrow \beta^n$ of $\mathbb{Z}$ on X. The spectrum of this representation is equal to the spectrum of β as an operator on X.

Theorem 4.9.: The following conditions are equivalent:

(1) $sp(\alpha)$ is compact.

(2) The representation α is uniformly continuous, i.e. $||id-\alpha_t|| \longrightarrow 0$ as $t \longrightarrow 0$, when id is the identity automorphism.

We can apply Corollary 4.8 to a *-automorphism Ψ in $Aut(\mathcal{B},\mathcal{a})$ and use Proposition 4.7 to conclude:

Corollary 4.10.: Let Ψ, in $Aut(\mathcal{B},\mathcal{a})$, be implemented by (I,φ), then the spectrum of Ψ as an operator on $\mathcal{B}(\mathcal{a},G)$ is $\overline{\cup_{U \in \bar{G}} \sigma(\varphi(U))}$. ###

Remark: Let $t \longrightarrow \alpha_t$ be a homomorphism from $\mathbb{R}$ into $Aut(\mathcal{B},\mathcal{a})$, implemented by (I,φ_t), where $\varphi_t(U) = \exp(itC(U))$ for every U in $\bar{G}$, and $C(U)$ are operators in $\mathcal{a}$ satisfying $\mathrm{Sup}_{U \in \bar{G}}||C(U)|| < \infty$. Then, by Theorems 4.6 and 4.9, the homomorphism $t \longrightarrow \alpha_t$ is uniformly continuous.

By Theorem VII.1.2 of [11] there is a bounded operator η on $\mathcal{B}(\mathcal{a},G)$ satisfying $\exp(it\,\eta) = \alpha_t$, and $\eta = \lim_{t \to 0} \frac{1}{t}(\alpha_t - id)$, where the limit is in the norm topology.

Derivations.

Let us denote by $D(g+g^*,\mathcal{a})$ the set of all skew adjoint derivations on $g+g^*$ vanishing on $\mathcal{a}$. i.e. a map δ from $g+g^*$ into $g+g^*$ is in $D(g+g^*,\mathcal{a})$ if:

(1) δ is linear.

(2) $\delta(TS) = T\,\delta(S) + \delta(T)S$ for T and S in $g+g^*$.

(3) $\delta(A) = 0$ for A in $\mathcal{a}$.

(4) $\delta(T^*) = -\delta(T)^*$ for T in $g+g^*$.

Lemma 4.11.: Let δ be in $D(g+g^*,\mathcal{a})$. Then there is a map C from $\bar{G}$ into $\mathcal{a}$

such that for each U and V in $\bar{G}$: (1) $\delta(U) = C(U)U$.

(2) $C(U)$ is self adjoint.

(3) $C(UV) = C(U)+UC(V)U^*$

We will say that δ is implemented by the map C. Conversely, if $C:\bar{G} \longrightarrow \mathcal{A}$ satisfies (2),(3) and $C(I) = 0$, then the map δ defined by $\delta(T) = \delta(\sum A_U U) = \sum A_U C(U)U$ (for $T = \sum A_U U$ in $g+g^*$) is in $D(g+g^*,\mathcal{A})$.

<u>Proof:</u> Let A be in $\mathcal{A}$ and U in $\bar{G}$, then

$0 = \delta(U^*AU) = \delta(U^*)AU+U^* \delta(AU) = \delta(U^*)AU+U^* \delta(A)U+U^*A \delta(U)$.

For $A = I$ we get $0 = \delta(U^*)U+U^* \delta(U)$; so that

$$\delta(U^*) + -U^* \delta(U)U^* \qquad (*).$$

Hence $0 = -U^* \delta(U)U^*AU+U^*A \delta(U)$ and $\delta(U)U^*AU = A \delta(U)$.

Thus $\delta(U)U^*$ commutes with each operator A in $\mathcal{A}$, hence $\delta(U)U^*$ is in $\mathcal{A}$ and we denote it by $C(U)$.

We have $\delta(U) = C(U)U$ and , since δ is skew adjoint,

$\delta(U^*) = -\delta(U)^* = -(C(U)U)^* = - U^*C(U)^*$; thus $C(U^*)U^* = -U^*C(U)^*$.

Using (*) we get : $C(U^*)U^* = \delta(U^*) = -U^* \delta(U)U^* = -U^*C(U)$. Hence $U^*C(U) = U^*C(U)^*$ and $C(U) = C(U)^*$, which proves (2).

Part (3) results from:

$C(UV)UV = \delta(UV) = \delta(U)V+U \delta(V) = C(U)UV+UC(V)V$.

To prove the last statement of the lemma it is enough, by linearity of the map δ, to prove that for each A,B in $\mathcal{A}$ and U,V in $\bar{G}$,

$\delta(AUBV) = \delta(AU)BV+AU \delta(BV)$ and $\delta((AU)^*) = -\delta(AU)^*$.

In fact, $\delta(AUBV) = \delta(AUBU^*UV) = AUBU^*C(UV)UV = AUBU^*C(U)UV + AUBU^*UC(V)U^*UV = AC(U)UBV + AUBC(V)V = \delta(AU)BV + AU \delta(BV)$.

Since $0 = C(I) = C(U^*) + U^*C(U)U$ $C(U^*) = -U^*C(U)U$;

hence $\delta((AU)^*) = \delta(U^*A^*UU^*) = U^*A^*UC(U^*)U^* = C(U^*)U^*A^* = U^*C(U)A^* =$

$= - (AC(U)U)^* = - \delta(AU)^*$. ###

The preceding lemma applies also to every skew adjoint derivation δ of $\mathcal{B}(\mathcal{A},G)$ that vanishes on $\mathcal{A}$. Thus every such derivation maps $g+g^*$ into $g+g^*$ and, hence, is an extension of a derivation in $D(g+g^*,\mathcal{A})$. We denote by $D(\mathcal{B},\mathcal{a})$ the set of all skew adjoint derivations on $\mathcal{B}(\mathcal{a},G)$ that vanish on $\mathcal{a}$. Every such derivation is bounded (since every derivation of a C*-algebra is bounded, by [47, Lemma 4.1.3]), and, as noted above, is an extension of a (bounded) derivation in $D(g+g^*,\mathcal{a})$.

On the other hand, if δ, in $D(g+g^*,\mathcal{a})$, is bounded then it can be extended, in a unique way, to a map δ' from $\mathcal{B}(\mathcal{a},G)$ into itself. Since δ is a skew adjoint derivation and $g+g^*$ is dense in $\mathcal{B}(\mathcal{a},G)$, δ' is a skew adjoint derivation on $\mathcal{B}(\mathcal{a},G)$ by continuity. Thus δ is in $D(\mathcal{B},\mathcal{a})$.

Therefore there is one-to-one correspondence between $D(\mathcal{B},\mathcal{a})$ and the set of all bonded derivations in $D(g+g^*,\mathcal{a})$. We will sometimes refer to the latter set as $D(\mathcal{B},\mathcal{a})$.

Definition: A derivation δ in $D(g+g^*,\mathcal{a})$ will be called inner if there is an operator D in $B(H)$ such that $\delta(T) = DT-TD$ for each T in $g+g^*$. In this case we write $\delta = ad(D)$.

Lemma 4.12.:

(1) δ, in $D(g+g^*,\mathcal{a})$, is inner (i.e. $\delta = ad(D)$), then D is in $\mathcal{a}$ and we can choose D such that $D = D^*$.

(2) The set of all inner derivations in $D(g+g^*,\mathcal{a})$ is $D(\mathcal{B},\mathcal{a})$.

Proof:

(1) Let δ be $ad(D)$. For each A in $\mathcal{a}$, $\delta(A) = 0$ hence $DA = AD$; so that $D \in \mathcal{a}' = \mathcal{a}$.

Let D_1 be $\frac{1}{2}(D+D^*)$, then D_1 is self adjoint and, to complete the proof of part (1), it will suffice to show that $ad(D_1)$ equals $ad(D)$ $(=\delta)$.

In fact, if T is in $g+g^*$ then $DT - TD = \delta(T) = -\delta(T^*)^* = -(DT^* - T^*D)^* = -TD^* + D^*T$ hence $ad(D_1)(T) = D_1T - DT_1 = \frac{1}{2}(DT+D^*T-TD-TD^*) = DT - TD = ad(D)(T)$.

(2) If δ is inner then, clearly, it is bounded. Let δ be a bounded derivation and extend it to a derivation on $\mathcal{B}(\mathcal{a},G)$.

It is known (see [47, Corollary 4.17]) that for a derivation δ on a C^*-algebra $\mathcal{C}$, there is an element A in the weak-operator closure of $\mathcal{C}$ satisfying $\delta = \mathrm{ad}(A)$. The application of this result to the algebra $\mathcal{B}(\mathcal{a},G)$ completes the proof. ###

If A is in $\mathcal{a}$ we let $\exp(A)$ be the sum of the norm convergent series

$$\sum_{n=0}^{\infty} \frac{1}{n!} A^n.$$

For δ in $D(b,\mathcal{a})$ we can define $\exp(i\delta)$ by the series $\sum_{n=0}^{\infty} \frac{1}{n!}(i\delta)^n$. Since δ is bounded, the series is convergent and defines a linear bounded operator on $\mathcal{B}(\mathcal{a},G)$.

For δ in $D(g+g^*),\mathcal{a})$, A in $\mathcal{a}$, and U in $\bar{G}$, $\delta(AU) = C(U)AU$ for some $C(U)$ in $\mathcal{a}$ (see Lemma 4.11); hence δ is bounded on $\mathcal{a}U$ (a Banach space) and $\exp(i\delta)$ is defined on $\mathcal{a}U$ by $(\exp(i\delta))(AU) = \sum_{n=0}^{\infty} \frac{1}{n!}(i\delta)^n(AU)$. Hence

$$(\exp(i\delta))(AU) = (\exp(iC(U)))AU.$$

We can, therefore, define $\exp(i\delta)$ on $g+g^*$ by:

$$(\exp(i\delta))(A_0+A_1U_1+\dots+A_nU_n) = A_0+(\exp(iC(U_1)))A_1U_1+\dots+(\exp(iC(U_n)))A_nU_n.$$

Lemma 4.13.: Let A and B be operators in $\mathcal{a}$, and U in $\bar{G}$. Then:

(1) $\exp(A+B) = (\exp A)(\exp B)$

(2) $\exp(U^*AU) = U^*(\exp A)U$.

If $\delta = \mathrm{ad}(D)$ for some D in $\mathcal{a}$, and T is in $\mathcal{B}(\mathcal{a},G)$, then

(3) $(\exp(i\delta))(T) = (\exp(iD))T(\exp(-iD))$.

Proof: (1) The algebra $\mathcal{a}$ is $*$-isomorphic to the algebra $C(X)$ of the continuous functions on some compact Hausdorff space X. For every f and g in $C(X)$ $\exp(f+g) = (\exp f)(\exp g)$ since equality holds pointwise on X. Hence (1) foloows.

(2) If φ is a $*$-representation of a C^*-algebra $\mathcal{C}$ and A is a self adjoint (or normal) element in $\mathcal{C}$ then for each continuous funciton f defined on the spectrum of A, $\varphi(f(A)) = f(\varphi(A))$. In the present case, let $\varphi(A)$ be U^*AU (a $*$-representation of the algebra $\mathcal{a}$) and f be $t \longrightarrow \exp(t)$;

hence $\exp(U^*AU) = U^*(\exp A)U$.

(3) Since $\exp(i\delta)$ is bounded and linear, it will suffice to prove the assertion for $T = AU$ where A is in $\mathcal{a}$ and U is in $\bar{G}$. As noted above, $(\exp(i\delta))(AU) = (\exp(iC(U)))AU$ where δ is implemented by the map C from $\bar{G}$ into $\mathcal{a}$. For $\delta = \mathrm{ad}(D)$ we have $\delta(U) = DU - UD = (D-UDU^*)U$; so that $C(U) = D-UDU^*$. Therefore

$$\begin{aligned}(\exp(i\delta))(AU) &= (\exp(iD))(\exp(-iUDU^*))AU \\ &= (\exp(iD))U(\exp(-iD))U^*AU \\ &= (\exp(iD))AU(\exp(-iD)).\end{aligned}$$ ###

Since we will use the conclusion of Theorem 3.13, we will assume, throughout the rest of this chapter, that conditions (1) and (2) of that theorem are satisfied; i.e. there is a faithful state g of $\mathcal{B}(\mathcal{a},G)$ satisfying, for A in $\mathcal{a}$ and U in $G\setminus\{I\}$: (1) $g(AU) = 0$ and (2) $g(U^*A^*AU) = g(A^*A)$.

Lemma 4.14.: For every δ in $D(\mathcal{g}+\mathcal{g}^*,\mathcal{a})$ the map $T \longrightarrow \exp(i\delta)(T)$, from $\mathcal{g}+\mathcal{g}^*$ into itself, can be extended, in only one way, to a *-automorphism in $\mathrm{Aut}(\mathcal{B},\mathcal{a})$. We denote the extension, again, by $\exp(i\delta)$ and write $\mathcal{E}$ for the map, so obtained, from $D(\mathcal{g}+\mathcal{g}^*,\mathcal{a})$ into $\mathrm{Aut}(\mathcal{B},\mathcal{a})$; i.e. $\mathcal{E}(\delta) = \exp(i\delta)$. For bounded derivation δ, this extension coincides with the sum of the convergent series $\sum_{n=0}^{\infty} \frac{1}{n!}(i\delta)^n$.

Proof: We have defined $\exp(i\delta)$ on $\mathcal{g}+\mathcal{g}^*$ by $(\exp(i\delta))(A_0+A_1U_1+\ldots+A_nU_n) = A_0 + (\exp(iC)(U_1)))A_1U_1 +\ldots+(\exp(iC(U_n)))A_nU_n$ where δ is implemented by the map $U \longrightarrow C(U)$, A_i are in $\mathcal{a}$, and U_i in $\bar{G}$. We will now fix δ and write Ψ for the map $T \longrightarrow \exp(i\delta)(T)$.

We, first, show that Ψ is multiplicative. For this, note that $C(UV) = C(U) + UC(V)U^*$ for U and V in $\bar{G}$ (see Lemma 4.11). Hence, for A,B in $\mathcal{a}$ and U, V in $\bar{G}$, : $\Psi(AUBV) = \Psi(AUBU^*UV) = AUBU^*(\exp(iC(UV)))UV = AUBU^*(\exp(iC(U)))U(\exp(iC(V)))U^*UV = \Psi(AU)\Psi(BV)$. By linearity of Ψ, this shows that Ψ is a homomorphism and, since $\Psi(U) = \exp(iC(U)))U$ is a unitary operator, Ψ is a *-homomorphism.

Let Ψ' be the map $T \longrightarrow \exp(-i\delta)(T)$, then, $\Psi'(AU) = (\exp(-iC(U)))AU$; so that $\Psi'(\Psi(AU)) = \Psi(\Psi'(AU)) = AU$ for $A \in \mathcal{a}$, $U \in \bar{G}$. Hence, by linearity $\Psi' = \Psi^{-1}$; so that Ψ is a $*$-automorphism on $\mathcal{g}+\mathcal{g}^*$.

Using Corollary 3.14 (where Ψ is implemented by $\Psi(U) = \exp(iC(U))$) we can extend Ψ, in a unique way, to a $*$-automorphism in $\mathrm{Aut}(\mathcal{B},\mathcal{a})$.

If δ is in $D(\mathcal{B},\mathcal{a})$, let Ψ be defined as above and let Ψ' be the sum of the convergent series $\sum_{n=0}^{\infty} \frac{1}{n!} (i\delta)^n$. Then, for A in $\mathcal{a}$ and U in $\bar{G}$, $\Psi'(AU) = \sum_{n=0}^{\infty} \frac{1}{n!}(i\delta)^n(AU) = (\exp(iC(U)))AU = \Psi(U)$. Since both Ψ and Ψ' are linear and continuous, $\Psi = \Psi'$. ###

Lemma 4.15.: A $*$-automorphism Ψ on $\mathcal{B}(\mathcal{a},G)$ is in $E(D(\mathcal{B},\mathcal{a}))$ if and only if Ψ is implemented by some unitary operator V in $\mathcal{a}$.

Proof: a) Let Ψ be $E(\delta)$ for some δ in $D(\mathcal{B},\mathcal{a})$. By Lemma 4.12, $\delta = \mathrm{ad}(D)$ for some self adjoint operator D in $\mathcal{a}$. Let V be the operator $\exp(iD)$ then, by Lemma 4.13 (3), Ψ is implemented by V.

b) Assume now that Ψ is implemented by V i.e. $\Psi(T) = VTV^*$ for T in $\mathcal{B}(\mathcal{a},G)$. There is a self adjoint operator D in $\mathcal{a}$ such that $\exp(iD) = V$. Let δ be $\mathrm{ad}(D)$. Then by Lemma 4.13 (3), $\Psi = E(\delta)$. ###

Lemma 4.16.: Let Ψ be in $\mathrm{Aut}(\mathcal{B},\mathcal{a})$ such that either $||\mathrm{id} - \Psi|| < 2$, or $\nu(\mathrm{id} - \Psi) < \sqrt{3}$, then Ψ is in $E(D(\mathcal{B},\mathcal{a}))$. Here, id is the identity automorphism of $\mathcal{B}(\mathcal{a},G)$ and $\nu(T)$ is the spectral radius of T.

Proof: This is Theorem 8.7.7 of [42]. ###

Corollary 4.17.: Let $\mathcal{g}_1$ and $\mathcal{g}_2$ be irreducible triangular algebras generated by $\mathcal{a}_1, G_1$ and $\mathcal{a}_2, G_2$ respectively, and assume that $\mathcal{g}_1$ satisfies conditions (1) and (2) of Theorem 3.13. Let Ψ and Ψ' be $*$-isomorphisms from $\mathcal{g}_1$ onto $\mathcal{g}_2$ such that Ψ is implemented by a unitary transformation W and $||\Psi - \Psi'|| < 2$. Then Ψ' is implemented by a unitary transformation. In particular, the set $E(D(\mathcal{B},\mathcal{a}))$ (in $\mathrm{Aut}(\mathcal{B},\mathcal{a})$) is open.

Proof.: Since Ψ is implemented by a unitary operator, Ψ is an isometry;

so that

$$||id - \Psi^{-1}\cdot\Psi'|| = ||\Psi - \Psi'|| < 2.$$

Thus $\Psi^{-1}\cdot\Psi'$ extends to a *-automorphism Ψ_0 of $\mathcal{B}(\mathcal{a},G)$; and $||id - \Psi_0|| < 2$. If F is a non zero projection in $\mathcal{a}_1$ then $\Psi_0(F)$ is another non zero projection in $\mathcal{a}_1$ and $||F - \Psi_0(F)|| < 2$, hence $\Psi_0(F) = F$. Thus $\Psi_0(A) = A$ for each A in $\mathcal{a}_1$ and, by Lemma 4.16, Ψ_0 is in $\mathcal{E}(D(\mathcal{B}_1,\mathcal{a}_1))$; and, from Lemma 4.15, Ψ_0 is implemented by a unitary operator V in $\mathcal{a}_1$. Since Ψ is implemented by W; we have,

$$\Psi'(T) = \Psi(\Psi_0(T)) = W^*V^*TVW,$$

for each T in $\mathcal{g}_1$. Thus Ψ' is implemented by VW. Note that if Ψ is in $\mathcal{E}(D(\mathcal{B},\mathcal{a}))$ then, by Lemma 4.15, it is implemented by a unitary operator W in $\mathcal{a}$ hence, by what we have proved to this point, each *-automorphism Ψ' satisfying $||\Psi - \Psi'|| < 2$ is also implemented by some unitary operator in $\mathcal{a}$ and thus, by Lemma 4.15, is in $\mathcal{E}(D(\mathcal{B},\mathcal{a}))$.
This proves the last assertion of the corollary, ###

Let $Aut_0(\mathcal{B},\mathcal{a})$ denote the connected component of the identity in $Aut(\mathcal{B},\mathcal{a})$ (in the norm topology).

Proposition 4.18.: $Aut_0(\mathcal{B},\mathcal{a}) = \mathcal{E}(D(\mathcal{B},\mathcal{a}))$.

Proof: Each Ψ in $\mathcal{E}(D(\mathcal{B},\mathcal{a}))$ is $\exp(i\delta)$ for some δ in $D(\mathcal{B},\mathcal{a})$. For every t in $\mathbb{R}$, $t\delta \in D(\mathcal{B},\mathcal{a})$ hence, by Lemma 4.14, $\exp(it\delta)$ $(=\beta(t))$ is in $Aut(\mathcal{B},\mathcal{a})$. Since δ is bounded, the map $t \longrightarrow \beta(t)$ is norm continuous and $\beta(0) = id$, $\beta(1) = \Psi$. Therefore Ψ lies in the connected component of the identity in $Aut(\mathcal{B},\mathcal{a})$.

Conversely, by Lemma 4.17, the set $\mathcal{E}(D(\mathcal{B},\mathcal{a}))$ is open. This set, by Lemma 4.15, is the set of all the *-automorphisms in $Aut(\mathcal{B},\mathcal{a})$ that are implemented by a unitary operator in $\mathcal{a}$; hence $\mathcal{E}(D(\mathcal{B},\mathcal{a}))$ is a subgroup of $Aut(\mathcal{B},\mathcal{a})$. Since it is open, it is also closed (in the norm topology).
Since $\mathcal{E}(D(\mathcal{B},\mathcal{a})) \subseteq Aut_0(\mathcal{B},\mathcal{a})$ and $Aut_0(\mathcal{B},\mathcal{a})$ is connected, $\mathcal{E}(D(\mathcal{B},\mathcal{a})) = Aut_0(\mathcal{B},\mathcal{a})$. ###

Lemma 4.19.: Let δ, in $D(\mathcal{G}+\mathcal{G}^*,\mathcal{A})$, be implemented by a map C from $\bar{G}$ to $\mathcal{A}$. Let Ψ_t be $E(t\delta)$. Then Ψ_t is implemented by (I, φ_t) where $\varphi_t(U) = \exp(itC(U))$ for each t in $\mathbb{R}$; and the map $t \longrightarrow \Psi_t(T)$ is continuous for each T in $\mathcal{B}(\mathcal{A},G)$.

Proof: By the definition of $E(t\delta)$ (see Lemma 4.14) Ψ_t is implemented by (I,φ_t).

For the continuity, let T be AU with A in $\mathcal{A}$ and U in $\bar{G}$, then $||\Psi_t(AU)-\Psi_s(AU)||= ||A(\exp(itC(U)) - \exp(isC(U)))U|| \leq ||A|| \, ||\exp(itC(U)) -\exp(isC(U))|| \leq ||A|| \, |t-s| \, ||C(U)|| \longrightarrow 0$ (as $t \longrightarrow s$). This proves the continuity when T is in $\mathcal{G}+\mathcal{G}^*$. If T is in $\mathcal{B}(\mathcal{A}, G)$, there is a sequence $\{S_n\}$ in $\mathcal{G}+\mathcal{G}^*$ such that $||S_n-T|| \longrightarrow 0$ as $n \to \infty$. Since, for all t in $\mathbb{R}$,

$$||\Psi_t(T) - \Psi_t(S_n)|| = ||T - S_n||,$$

the function $t \longrightarrow \Psi_t(T)$ is a uniform limit of the continuous functions $t \longrightarrow \Psi_t(S_n)$ on $\mathbb{R}$. Hence $t \longrightarrow \Psi_t(T)$ is continuous for each T in $\mathcal{B}(\mathcal{A},G)$.

###

Theorem 4.20.: Let δ, in $D(\mathcal{G}+\mathcal{G}^*,\mathcal{A})$, be implemented by the map $U \longrightarrow C(U)$ with $\operatorname{Sup}_{U \in \bar{G}}||C(U)|| < \infty$, then δ is in $D(\mathcal{B},\mathcal{A})$.

Proof: Let Ψ_t be $\exp(it\delta)$, then by the remark following Corollary 4.10, there is a bounded operator η on $\mathcal{B}(\mathcal{A},G)$ such that $\exp(it\eta) = \Psi_t$ for t in $\mathbb{R}$. Since $\eta= \lim_{t\to 0} \frac{1}{t}(\Psi_t-\mathrm{id})$ we have, for A in $\mathcal{A}$ and U in $\bar{G}$:

$$\eta(AU) = \lim_{t\to 0} \frac{1}{t}A(\exp(itC(U))U-U)= A(\lim_{t\to 0} \frac{1}{t}(\exp(itC(U))-I))U=AC(U)U= \delta(AU)$$

where the limits are in the norm topology. Hence $\delta=\eta$ on $\mathcal{G}+\mathcal{G}^*$ and so δ is bounded on $\mathcal{G}+\mathcal{G}^*$, thus is in $D(\mathcal{B},\mathcal{A})$.

###

For a C*-algebra C, acting on a Hilbert space H, let $(C_+)^m$ denote the set of all oprators in $B(H)$ that are strong limits of monotone increasing nets of positive operators from C. Let $((C_+)^m)^-$ denotes its norm closure.

The next two results are Proposition 8.6.4 and Theorem 8.6.5 of [47].

Proposition 4.21.: If δ is a *-derivation of a C*-algebra C, define Ψ_t to be $\exp(t\delta)$ for each t in $\mathbb{R}$. Then $\Psi: t \longrightarrow \Psi_t$ is a uniformly continuous one parameter group of automorphisms of $\mathcal{C}$ and $\sigma(\delta) = i\ sp(\Psi)$.

Theorem 4.22.: Let δ be a *-derivation of a C*-algebra $\mathcal{C}$. Then the spectral radius of δ equals $||\delta||$ and there is an element h in $((\mathcal{C}_+)^m)^-$ such that $\delta = ad(ih)$ and $||\delta|| = ||h||$.

Applying these results we get:

Corollary 4.23.: Let δ , in $D(\mathcal{G}+\mathcal{G}^*,\mathcal{a})$, be implemented by the map $U \to C(U)$ from $\bar{G}$ to $\mathcal{a}$, such that $\mathrm{Sup}_{U \in \bar{G}} ||C(U)|| < \infty$. Then there is a self adjoint operator D in $\mathcal{a}$, such that $\delta = ad(D)$ and:

(1) $\sigma(\delta) = \overline{\cup_{U \in \bar{G}} \sigma(C(U))}$.

(2) $||D|| = ||\delta|| = \mathrm{Sup}_{U \in \bar{G}} ||C(U)||$.

Proof: By Theorem 4.20, δ is in $D(\mathcal{B},\mathcal{a})$, hence, by Lemma 4.12, there is a self adjoint operator D in $\mathcal{a}$ such that $\delta = ad(D)$. Let Ψ be the map $t \longrightarrow \exp(it\delta)$. Since δ is skew adjoint, the derivation $i\delta$ is a *-derivation (i.e. a self adjoint derivation) hence, applying Proposition 4.21 we get $\sigma(i\delta) = i\ sp(\Psi)$; so that $\sigma(\delta) = sp(\Psi)$. Since δ is implemented by the map $U \longrightarrow C(U)$; by Lemma 4.19, $\exp(it\delta)$ is implemented by the map $U \longrightarrow \exp(itC(U))$.

Hence we can apply Theorem 4.6 to get

$$\sigma(\delta) = sp(\Psi) = \overline{\cup_{U \in \bar{G}} \sigma(C(U))}.$$

For the second part, note that $||C(U)|| = \mathrm{Sup}\{|t| : t \in \sigma(C(U))\}$ hence $\mathrm{Sup}\ \{|t|: t \in \sigma(\delta)\} = \mathrm{Sup}\{|t|: t \in \sigma(CU)),\ U \in \bar{G}\} = \mathrm{Sup}\{||C(U)||: U \in \bar{G}\}$.
By Theorem 4.22, the spectral radius of δ equals $||\delta||$. Hence

$$||\delta|| = \mathrm{Sup}_{U \in \bar{G}} ||C(U))||.$$

By Theorem 4.22, there is a self adjoint operator D such that $\delta = ad(D)$ and $||\delta|| = ||D||$: by Lemma 4.12, D is in $\mathcal{a}$. ###

Thus far, in this chapter, we have dealt only with *-automorphisms. The following theorem discusses an automorphism that is not necessarily a *-automorphism.

Theorem 4.24.: Let Ψ be an automorphism (not necessarily a *-automorphism) of $\mathcal{G}$ leaving each A in $\mathcal{a}$ fixed. Assume also that Ψ is implemented by (I,φ) and that there are positive numbers N,M such that for each U in G, $NI \leq |\varphi(U)| \leq MI$. Then Ψ can be extended to a bounded automorphism Ψ' on $\mathcal{B}(\mathcal{a},G)$ and $\Psi' = \Psi_1' \cdot \Psi_2'$ where:

(1) Ψ_2' is in $\mathrm{Aut}(\mathcal{B},\mathcal{a})$.

(2) $\Psi_1'(T) = RTR^{-1}$ for each T in $\mathcal{B}(\mathcal{a},G)$, where R is some invertible operator in $\mathcal{a}$.

Proof: Using the polar decomopsition, we can define two maps, φ_1 and φ_2, from G into $\mathcal{a}$ such that $\varphi_1(U) = |\varphi(U)|$, $\varphi(U) = \varphi_1(U)\,\varphi_2(U)$, and $\varphi_2(U)$ is a partial isometry for each U in G. Since $\varphi(U)$ is invertible, $\varphi_2(U)$ is, in fact, a unitary operator. For each U,V in G we have:

$$\varphi(UV) = \varphi(U)U\,\varphi(V)U^* = \varphi_1(U)U\,\varphi_1(V)U^*\,\varphi_2(U)U\,\varphi_2(V)U^*.$$

Hence, by the uniqueness of the polar decomposition, we get:

$$\varphi_i(UV) = \varphi_i(U)U\,\varphi_i(V)U^* \qquad i = 1,2 \qquad \text{for } U,V \text{ in } G.$$

Therefore, by the discussion preceding Lemma 3.4, (I,φ_1) and (I,φ_2) implement two homomorphisms, Ψ_1 and Ψ_2, on $\mathcal{G}$, and $\Psi = \Psi_1 \cdot \Psi_2$. Since $\varphi_2(U)$ is a unitary operator for each U in G, the homomorphism implemented by (I, φ_2^*), where $\varphi_2^*(U) = \varphi_2(U)^*$, is the inverse homomorphism of Ψ_2. Hence Ψ_2 is an automorphism and so is $\Psi_1 (= \Psi \cdot \Psi_2^{-1})$. Since $\varphi_2(U)$ is a unitary operator for each U in G, Ψ_2 can be extended, by Corollary 3.14, to a *-automorphism Ψ_2' in $\mathrm{Aut}(\mathcal{B},\mathcal{a})$.

Each operator $|\varphi(U)|$ (in $\mathcal{a}$) has its spectrum contained in the interval $[N,M]$, $N > 0$. Thus, using Lemma 1 of [22], we can define the operator $\log|\varphi(U)|$ in $\mathcal{a}$, and we have:

$$\log|\varphi(UV)| = \log|\varphi(U)| + \log U|\varphi(V)|U^* =$$
$$\log|\varphi(U)| + U(\log|\varphi(V)|)U^*.$$

Let $C(U)$ be $\log|\varphi(U)|$, then:

$$C(UV) = C(U) + UC(V)U^* \qquad (*)$$

for U and V in G.

The spectrum of $C(U)$ lies in the interval $[\log(N), \log(M)]$, hence $C(U)$

is self adjoint (it lies in $\mathcal{A}$ hence it is normal) and

$$\operatorname{Sup}_{U \in \bar{G}} ||C(U)|| < \infty.$$

For U in G, define $C(U^*) = -U^*C(U)U$, then for each U in $\bar{G}$, C(U) is self adjoint and

$$\operatorname{Sup}_{U \in \bar{G}} ||C(U)|| < \infty \qquad (**).$$

Note also that (*) holds for U,V in $\bar{G}$. We prove this for the case where $U \in G$, $V \in G^*$ and $UV \in G$. The other possible cases are handled similarly. Under these assumptions, it follows from (*) that

$$C(U) = C(UV) + UVC(V^*)V^*U^*.$$

Since $C(V) = -VC(V^*)V^*$, $C(U) = C(UV) - UC(V)U^*$.

Hence $C(UV) = C(U) + UC(V)U^*$.

Define a map $\delta\colon \mathcal{g}+\mathcal{g}^* \longrightarrow \mathcal{g}+\mathcal{g}^*$ by $\delta(\sum A_U U) = \sum C(U)A_U U$. To see that δ is a derivation, note that, using (*), we have

$\delta(AUBV) = AUBU^*C(UV)UV = AUBU^*C(U)UV +$

$\delta(AUBU^*UC(V)U^*UV = AC(U)UBV + AUBC(V)V = \delta(AU)BV + AU\,\delta(BV)$ for A,B in $\mathcal{A}$ and U,V in $\bar{G}$. Since δ is linear, this implies that δ is a derivation on $\mathcal{g}+\mathcal{g}^*$. Since $C(I) = \log|\varphi(I)| = \log(I) = 0$, $\delta(A) = 0$ for each A in $\mathcal{A}$.

Also $\delta((AU)^*) = \delta(U^*A^*) = U^*A^*UC(U^*)U^* = -\, U^*A^*UU^*C(U)UU^*$

$= -U^*A^*C(U) = -\,\delta(AU)^*$; so that δ is in $D(\mathcal{g}+\mathcal{g}^*,\mathcal{A})$.

Moreover, by (**), the hypothesis of Corollary 4.23 is satisfied hence there is a self adjoint operator D in $\mathcal{A}$ such that $\delta = \operatorname{ad}(D)$; so that $C(U)U = DU - UD$ and $C(U) = D-UDU^*$.

Let the operator R, in $\mathcal{A}$, be exp(D) then $R^{-1} = \exp(-D)$ and $|\varphi(U)| = \exp(C(U)) = \exp(D-UDU^*) = (\exp(D))U(\exp(-D))U^* = RUR^{-1}U^*$ for each U in $\bar{G}$.

Define a map τ on $\mathcal{g}$ by $\tau(T) = RTR^{-1}$ then, for A in $\mathcal{A}$, U in $\bar{G}$, we have:

$$\tau(AU) = RAUR^{-1} = ARUR^{-1} = A|\varphi(U)|U = \Psi_1(AU) \quad ;$$

so that, by linearity, $\tau = \Psi_1$ on $\mathcal{g}$. Since τ can be extended to an automorphism Ψ_1' on $\mathcal{B}(\mathcal{A},G)$ so can Ψ_1. ###

Example 4.25. : Let $\mathcal{A}$ be the multiplication algebra of $L^2([-\Pi,\Pi])$ (i.e. $\mathcal{A} \cong L^\infty([-\Pi,\Pi])$). Let G be the semigroup generated by a single unitary operator U acting ergodically on $\mathcal{A}$. By Lemma 1.7, the algebra $\mathcal{G}$ generated by $\mathcal{A}$ and G is an irreducible triangular algebra.

Fix t in $[-\Pi,\Pi]$ and, for each integer k, let $\varphi_t(U^k)$ be $e^{itk}I$. Define a linear map α_t on $\mathcal{G}$ by: $\alpha_t(A_0+A_1U+\dots+A_nU^n) = A_0 + e^{it}A_1U+\dots+e^{itn}A_nU^n$ where A_i is in $\mathcal{A}$.

For A,B in $\mathcal{A}$, we have

$$\alpha_t(AU^kBU^m) = e^{it(k+m)}AU^kBU^m = \alpha_t(AU^k)\ \alpha_t(BU^m) \ ;$$

so that, by linearity, α_t is a homomorphism on $\mathcal{G}$. Note that Λ is $\mathbb{T}$, in the present case.

We assume now that U arises from a measure-preserving transformation on $[-\Pi,\Pi]$ (with the Lebesgue measure dt). Let g be a state on $\mathcal{A}$ defined by $g(f) = \frac{1}{2\Pi}\int_{-\Pi}^{\Pi} f(t)dt$; then g is a faithful G-invariant state on $\mathcal{A}$.

Since $\bar{G}$ is amenable (hence Φ is faithful), the remark preceding Corollary 3.14 implies that the conditions of that Corollary are satisfied.

Since α_t is implemented by φ_t, and $\varphi_t(U^k)$ is a unitary operator for each positive integer k, we can apply Corollary 3.14 and extend α_t to a *-automorphism on the C*-algebra $\mathcal{B}(\mathcal{A},G)$. Denote the extension, again, by α_t.

The map $t \longrightarrow \alpha_t(T)$ is continuous if T is of the form AU^k $(A \in \mathcal{A}, k \in \mathbb{Z})$ because $\alpha_t(AU^k) = e^{itk}AU^k$. Hence the map is continuous for each T in $\mathcal{G}+\mathcal{G}^*$. As in the proof of Lemma 4.19, the map $t \longrightarrow \alpha_t(T)$ is continuous for each T in $\mathcal{B}(\mathcal{A},G)$..

Thus, to each μ in $M([-\Pi,\Pi])$, we can associate an operator $\alpha_\mu(T)$ in $\mathcal{B}(\mathcal{A},G)$, (See Corollary 4.2 (1).) For every functional g in the dual of $\mathcal{B}(\mathcal{A},G)$, we have: $g(\alpha_\mu(T)) = \int_{-\Pi}^{\Pi} g(\alpha_t(T))d\mu(t)$.

Hence $\quad |g(\alpha_\mu(T))| \leq ||g||\ ||T||\ ||\mu||$; so that

$$||\alpha_\mu(T)|| \leq ||T|| \quad .$$

A sequence $\{K_n\}$ of functions in $L^1([-\Pi,\Pi])$ is called an <u>approximate</u>

<u>identity</u> if they satisfy:

(1) $K_n \geq 0$.

(2) $\frac{1}{2\Pi}\int_{-\Pi}^{\Pi} K_n(t)dt = 1$.

(3) If J is an open interval about $t = 0$ then

$$\lim_{n\to\infty} \operatorname{Sup}_{t\notin J} K_n(t) = 0.$$

Such a sequence is called an approximate identity since for every function f in $L^1([-\Pi,\Pi])$,

$$||K_n*f - f||_1 \underset{n}{\to} 0 .$$

We will now prove that $\alpha_{K_n}(T) \underset{n}{\longrightarrow} T$ in norm where K_n is an approximate identity. For This, let $\delta > 0$ and write, for each bounded linear functional g on $\mathcal{B}(\mathcal{a},G)$:

$$\begin{aligned} g(\alpha_{K_n}(T) - T) &= \frac{1}{2\Pi}\int_{-\Pi}^{\Pi} g(\alpha_t(T))K_n(t)dt - g(T) \\ &= \frac{1}{2\Pi}\int_{-\Pi}^{\Pi}(g(\alpha_t(T)) - g(T))K_n(t)dt \\ &= \frac{1}{2\Pi}\int_{-\delta}^{\delta}(g(\alpha_t(T)) - g(T))K_n(t)dt \\ &\qquad + \frac{1}{2\Pi}\int_{|t|\geq\delta}(g(\alpha_t(T)) - g(T))K_n(t)dt. \end{aligned}$$

Therefore

$$\begin{aligned} |g(\alpha_{K_n}(T) - T| &\leq \operatorname{Sup}_{-\delta\leq t\leq\delta}|g(\alpha_t(T)) - g(T)| \\ &+ 2||g||\ ||T||\operatorname{Sup}_{|t|\geq\delta}K_n(t) \leq ||g||(\operatorname{Sup}_{-\delta\leq t\leq\delta}||\alpha_t(T) - T|| \\ &+ 2||T||\operatorname{Sup}_{|t|\geq\delta}K_n(t)). \end{aligned}$$

Fix $\varepsilon > 0$. Since $t \longrightarrow \alpha_t(T)$ is continuous at $t = 0$ and $\alpha_0(T) = T$, we can choose $\delta(=\delta(\varepsilon))$ such that $\operatorname{Sup}_{-\delta\leq t\leq\delta}||\alpha_t(T) - T|| \leq \frac{1}{2}\varepsilon$.

Since $\lim_{n\to\infty} \operatorname{Sup}_{|t|\geq\delta}K_n(t) = 0$, there is an integer N such that for each $n \geq N$

$$\operatorname{Sup}_{|t|\geq\delta}K_n(t) \leq \tfrac{1}{4}\varepsilon||T|| \qquad \text{and thus}$$

$$|g(\alpha_{K_n}(T) - T)| \leq \varepsilon\ ||g|| .$$

Hence $$||\alpha_{K_n}(T) - T|| \underset{n}{\longrightarrow} 0 .$$

We can choose K_n to be the trigonometric polynomial (the classical

Fejer's Kernel) $\sum_{k=-n}^{n} a_n(k)e^{ikt}$, where $a_n(k) = 1-\frac{k}{n}$.

Then K_n is an approximate identity and $K_n(t) = \frac{1}{n}(\frac{\operatorname{Sin} 1/2nt}{\operatorname{Sin} 1/2t})^2$ (see [28, pp. 16-17]).

With this choice of K_n we get:

$$\alpha_{K_n}(T) = \sum_{k=-n}^{n} a_n(-k)\ \Phi(TU^{-k})U^k \qquad (*).$$

To prove this, let us, first, define a linear bounded map from $\mathcal{B}(\mathcal{a},G)$ onto by $\Phi_k(T) = \Phi(TU^{-k})$ $k \in \mathbb{Z}$ and observe that $\Phi_k \cdot \alpha_t = e^{itk}\ \Phi_k$ for each k in $\mathbb{Z}$ and t in $[-\Pi,\Pi]$. (This last equality is clear on $g+g^*$ and extends to $\mathcal{B}(\mathcal{a},G)$ by continuity of each side).

Let g be any bounded linear functional on $\mathcal{B}(\mathcal{a},G)$, we have:

$$g(\Phi_k(\alpha_{K_n}(T))) = \frac{1}{2\Pi}\int_{-\Pi}^{\Pi} g(\Phi_k(\alpha_t(T)))K_n(t)dt = \frac{1}{2\Pi}\int_{-\Pi}^{\Pi} e^{itk} g(\Phi_k(T))K_n(t)dt$$
$$= g\ (\Phi_k(T))a_n(-k),$$

where we let $a_n(k)$ be 0 if $|k| \geq n$.

Therefore, for every k, $\Phi_k(\alpha_{K_n}(T)) = a_n(-k)\ \Phi_k(T)$.

Since $\Phi_k(\sum_{j=-n}^{n} a_n(-j)\ \Phi(TU^{-j})U^j) = a_n(-k)\ \Phi_k(T)$ we conclude that:

$$\Phi_k(\alpha_{K_n}(T)) = \Phi_k(\sum_{j=-n}^{n} a_n(-j)\ \Phi(TU^{-j})U^j)$$

for each k in $\mathbb{Z}$.

Using Lemma 2.10 this completes the proof of (*).

In fact, the above argument shows that for every function f in $L^1([-\Pi,\Pi])$

$$\Phi(\alpha_f(T)U^{-k}) = \hat{f}(k)\ \Phi(TU^{-k}).$$

With the exception of the discussion related to K_n, the representation $t \longrightarrow \alpha_t$ applies to the case where $\Lambda = \mathbb{R}$. In this case $C(U^k) = kI$, and, applying Theorem 4.6 and Corollary 4.10, we have:

(1) $\operatorname{sp}(\alpha) = \mathbb{Z}$.

(2) For every t in $\mathbb{R}$, the spectrum of α_t as an operator on $\mathcal{B}(\mathcal{a},G)$ is the closure of the set $\{e^{ikt} : k \in \mathbb{Z}\}$.

Thus if e^{it} is not a root of 1, the spectrum is the whole circle group.

BIBLIOGRAPHY

1. W.B. Arveson, A density theorem for operator algebras. Duke Math J. 23 (1967), pp. 635-648.
2. W.B. Arveson, Analyticity in operator algebras. Amer. J. Math. 89 (1967), pp. 578-642.
3. W.B. Arveson, Operator algebras and measure preserving automorphisms. Acta Math. 118 (1967), pp. 95-109.
4. W.B. Arveson, Operator algebras and invariant subspaces. Ann. of Math. 100 (1974), pp. 433-532.
5. W.B. Arveson, On groups of automorphisms of operator algebras. J. Functional Anal. 15 (1974)
6. W.B. Arveson, Interpolation problems in nest algebras. J. Functional anal. 20 (1975), pp. 208-233.
7. W.B. Arveson and K.B. Josephson, Operator algebras and measure preserving automorphisms II. J. Functional Anal. 4 (1969), pp. 100-134.
8. G. Birkhoff, Lattice Theory. Amer. Math. Soc. Coll. Publ. v.5, third edition (1967).
9. E. Christensen, Derivations of nest algebras. Math. Ann. 229 (1977), pp. 155-161.
10. J.A. Deidens, Another description of nest algebras. Lecture Notes in Mathematics No. 693. Springer-Verlag, New York 1978, pp.77-86.
11. N. Dunford and J.T. Schwartz, Linear operators, Part 1: General theory. New York, N.Y., Interscience Publishers, Inc. 1958.
12. H.A. Dye, On groups of measure preserving transformations I. Amer. J. of Math. 81 (1959), pp. 119-159.
13. H.A. Dye, On groups of measure preserving transformations II. Amer. J. of Math. 85 (1963), pp. 551-576.
14. J.A. Erdos, Some results on triangular operator algebras, Amer. J. Math. 89 (1967), pp. 85-93.
15. J.A. Erdos, Unitary invariants for nests. Pacific J. Math. 23 (1967), pp. 229-256.
16. J.A. Erdos, Operators of finite rank in nest algebras. J. London Math. Soc. 43 (1968), pp. 391-397.
17. J.A. Erdos, An abstract characterization of next algebras. Quart. J. Math. Oxford 22 (1971), pp. 47-63.

18. J.A. Erdos, Some questions concerning triangular operator algebras. Proc. R. In. Acad. 74A (1974), pp. 223-232.

19. J.A. Erdos, On some ideals of nest algebras. Proc. London Math. Soc.

20. T. Fall, W.B. Arveson and P.S. Muhly, Perturbations of nest algebras. J. Operator Theory 1 (1979), pp. 137-150.

21. E. Følner, On groups with full Banach mean value. Math. Scand. 3 (1955), pp. 243-254.

22. T. Gardner, An invariance theorem for representations of Banach algebras. Proc. Amer. Math. Soc. 16 (1965), pp. 383.

23. F. Gilfeather and D.R. Larson, Nest subalgebras of von Neumann algebras. Advances in Math.

24. F. Gilfeather and D.R. Larson, Nest subalgebras of von Neumann algebras: commutants modulo compacts and distance estimates.

25. F. Gilfeather and D.R. Larson, Nest subalgebras of von Neumann algebras: commutants modulo the radical.

26. P.R. Halmos, Reflexive lattices of subspaces. J. London Math. Soc. (2) 4(1971), pp. 257-263.

27. P.R. Halmos, Measure theory. Springer-Verlag New York 1974.

28. K. Hoffman, Banach spaces of analytic functions, Prentice-Hall, Englewook Cliffs, N.J. 1962.

29. A. Hopenwasser, Isometries on irreducible triangular operator algebras. Math. Scand. 30 (1972), pp. 136-140.

30. A. Hopenwasser, The radical of a reflexive operator algebra. Pacific J. Math. 65 (1976), pp. 375-392.

31. A. Hopenwasser, The Cartan criterion fails for triangular subalgebras of a factor. Rocky Mountain J. Math. 9 (1979), pp. 441-445.

32. A. Hopenwasser, The equation Tx=y in a reflexive operator algebra. Indiana Univ. Math. J. 29 (1980), pp. 121-126.

33. R.V. Kadison and I.M. Singer, Triangular operator algebras. Amer. J. Math. 82 (1960), pp. 227-259.

34. E.C. Lance, Some properties of nest algebras. Proc. London Math. Soc. (3) 19 (1968), pp. 45-68.

35. E.C. Lance, Cohomology and perturbations of nest algebras.

36. D.R. Larson, On the structure of certain reflexive operator algebras. J. Functional Anal. 31 (1979), pp. 275-292.

37. W.E. Longstaff, Generators of nest algebras. Canadian J. Math.26 (1974), pp. 565-575.

38. W.E. Longstaff, Strongly reflexive lattices. J. London Math. Soc. (2) 11 (1975), pp. 491-498.

39. W.E. Longstaff, Generators of reflexive algebras. J. Australian Math. Soc. 20 (1975), pp. 159-164.

40. W.E. Longstaff, Operators of rank one in reflexive algebras. Canadian J. Math. 28 (1976), pp. 19-23.

41. F.J. Murray and J. von Neumann, On rings of operators. Ann. of Math. 37 (1936), pp. 116-229.

42. G.K. Pedersen,C*-algebras and their automorphism groups. Academic Press Inc. (London) 1979.

43. J.R. Ringrose, Algebraic isomorphisms between ordered bases. Amer. J. Math. 83 (1961), pp. 463-478.

44. R.J. Ringrose, On some algebras of operators. Proc. London Math. Soc. (3) 15 (1965), pp. 61-83.

45. J.R. Ringrose, On some algebras of operators II. Proc. London Math. Soc. (3) 16 (1966), pp. 385-402.

46. P. Rosental, Weakly closed maximal triangular algebras are hyperreducible. Proc. Amer. Math. Soc. 24 (1970), p.220.

47. S. Sakai, C*-algebras and W*-algebras. 'Ergebnisse der Mathematic' 60, Springer-Verlag, Berlin, Heidelberg, New York, 1971.

48. J.R. Schue, The structure of hyperreducible triangular algebras. Proc. Amer. Math. Soc. 15 (1964), pp. 766-772.

49. E. Størmer, Positive linear maps of operator algebras. Acta Math. 110 (1963), pp. 233-278.

School of Mathematical Sciences, Tel-Aviv University, Tel-Aviv, Israel

General instructions to authors for

PREPARING REPRODUCTION COPY FOR MEMOIRS

For more detailed instructions send for AMS booklet, "A Guide for Authors of Memoirs." Write to Editorial Offices, American Mathematical Society, P. O. Box 6248, Providence, R. I. 02940.

MEMOIRS are printed by photo-offset from camera copy fully prepared by the author. This means that, except for a reduction in size of 20 to 30%, the finished book will look exactly like the copy submitted. Thus the author will want to use a good quality typewriter with a new, medium-inked black ribbon, and submit clean copy on the appropriate model paper.

Model Paper, provided at no cost by the AMS, is paper marked with blue lines that confine the copy to the appropriate size. Author should specify, when ordering, whether typewriter to be used has PICA-size (10 characters to the inch) or ELITE-size type (12 characters to the inch).

Line Spacing – For best appearance, and economy, a typewriter equipped with a half-space ratchet – 12 notches to the inch – should be used. (This may be purchased and attached at small cost.) Three notches make the desired spacing, which is equivalent to 1-1/2 ordinary single spaces. Where copy has a great many subscripts and superscripts, however, double spacing should be used.

Special Characters may be filled in carefully freehand, using dense black ink, or INSTANT ("rub-on") LETTERING may be used. AMS has a sheet of several hundred most-used symbols and letters which may be purchased for $5.

Diagrams may be drawn in black ink either directly on the model sheet, or on a separate sheet and pasted with rubber cement into spaces left for them in the text. Ballpoint pen is *not* acceptable.

Page Headings (Running Heads) should be centered, in CAPITAL LETTERS (preferably), at the top of the page – just above the blue line and touching it.

LEFT-hand, EVEN-numbered pages should be headed with the AUTHOR'S NAME;

RIGHT-hand, ODD-numbered pages should be headed with the TITLE of the paper (in shortened form if necessary).

Exceptions: PAGE 1 and any other page that carries a display title require NO RUNNING HEADS.

Page Numbers should be at the top of the page, on the same line with the running heads.

LEFT-hand, EVEN numbers – flush with left margin;

RIGHT-hand, ODD numbers – flush with right margin.

Exceptions: PAGE 1 and any other page that carries a display title should have page number, centered below the text, on blue line provided.

FRONT MATTER PAGES should be numbered with Roman numerals (lower case), positioned below text in same manner as described above.

MEMOIRS FORMAT

It is suggested that the material be arranged in pages as indicated below.
Note: Starred items (*) are requirements of publication.

Front Matter (first pages in book, preceding main body of text).

Page i – *Title, *Author's name.

Page iii – Table of contents.

Page iv – *Abstract (at least 1 sentence and at most 300 words).

*1980 Mathematics Subject Classifications represent the primary and secondary subjects of the paper. For the classification scheme, see Annual Subject Indexes of MATHEMATICAL REVIEWS beginning in December 1978.

Key words and phrases, if desired. (A list which covers the content of the paper adequately enough to be useful for an information retrieval system.)

Page v, etc. – Preface, introduction, or any other matter not belonging in body of text.

Page 1 – Chapter Title (dropped 1 inch from top line, and centered).

Beginning of Text.

Footnotes: *Received by the editor date.
Support information – grants, credits, etc.

Last Page (at bottom) – Author's affiliation.

ABCDEFGHIJ–AMS–8987654